SIEMENS

NX EXERCISES

200 PRACTICE DRAWINGS

SACHIDANAND JHA

Dear Reader,

Thank you for choosing **SIEMENS NX EXERCISES** book. This book is part of a family of premium-quality CADIN360 books, all of which are written by Outstanding author who combine practical experience with a gift for teaching.

CADIN360 was founded in 2016. More than 3 years later, we're still committed to producing consistently exceptional books. With each of our titles, we're working hard to set a new standard for the industry. From the paper we print on, to the authors we work with, our goal is to bring you the best books available.

I hope you see all that reflected in these pages. I'd be very interested to hear your comments and get your feedback on how we're doing. Feel free to let me know what you think about this or any other CADIN360 book by sending me an email at contactus@cadin360.com.

If you think you've found a technical error in this book, please visit https://cadin360.com/contact-us/.
Customer feedback is critical to our efforts at CADIN360.

Best regards,

Sachidanand Jha
Founder & CEO, CADIN360

SIEMENS NX EXERCISES

Published by
CADIN360
cadin360.com

Preface

SIEMENS NX EXERCISES

- This book contain 200 CAD practice exercises and drawings.
- This book does not provide step by step tutorial to design 3D models.
- S.I Unit is used.
- Predominantly used Third Angle Projection.
- This book is for **NX** and Other Feature-Based Modeling Software such as Inventor, SolidWorks, Solid Edge, AutoCAD, PTC Creo etc.
- It is intended to provide Drafters, Designers and Engineers with enough 3D CAD exercises for practice on **NX**.
- It includes almost all types of exercises that are necessary to provide, clear, concise and systematic information required on industrial machine part drawings.
- Third Angle Projection is intentionally used to familiarize Drafters, Designers and Engineers in Third Angle Projection to meet the expectation of world wide Engineering drawing print.
- Clear and well drafted drawing help easy understanding of the design.
- This book is for Beginner, Intermediate and Advance CAD users.
- These exercises are from Basics to Advance level.
- Each exercises can be assigned and designed separately.
- No Exercise is a prerequisite for another. All dimensions are in mm.
- Note: Assume any missing dimensions.

EX-01

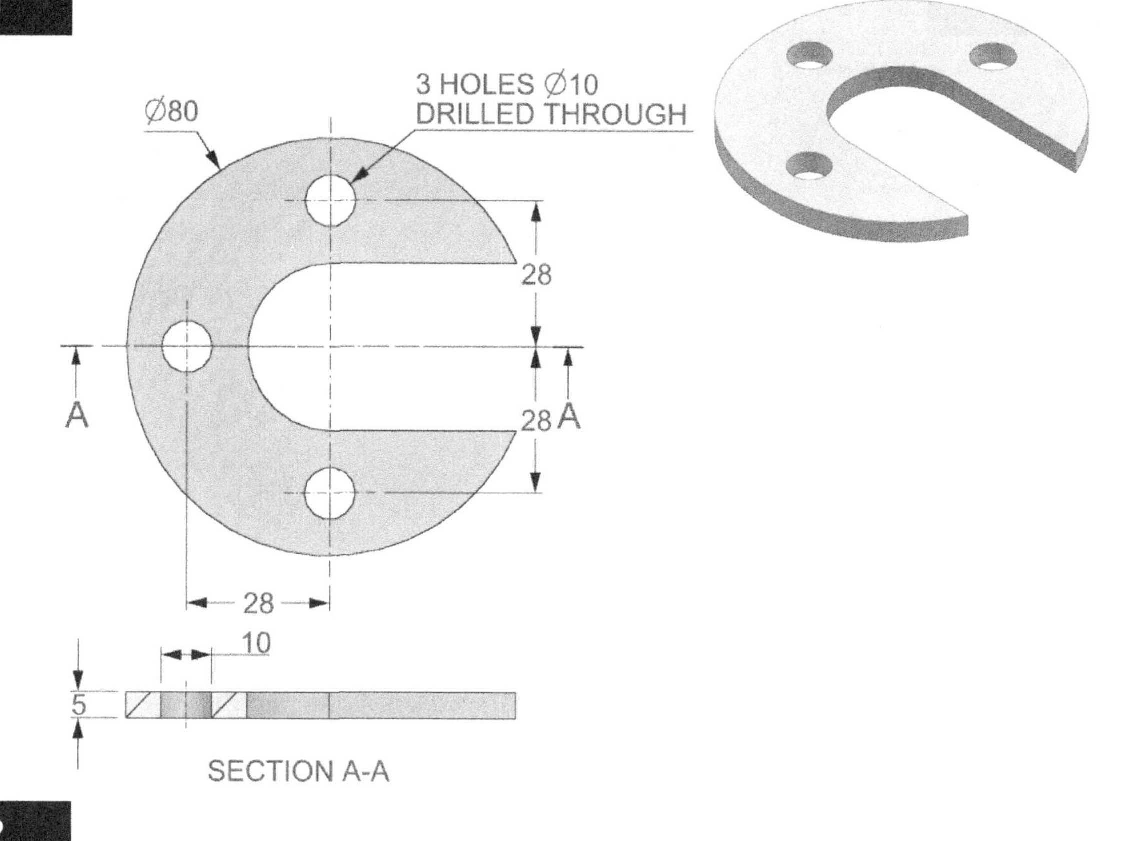

3 HOLES Ø10
DRILLED THROUGH
Ø80
28
28
A
A
28
10
5
SECTION A-A

EX-02

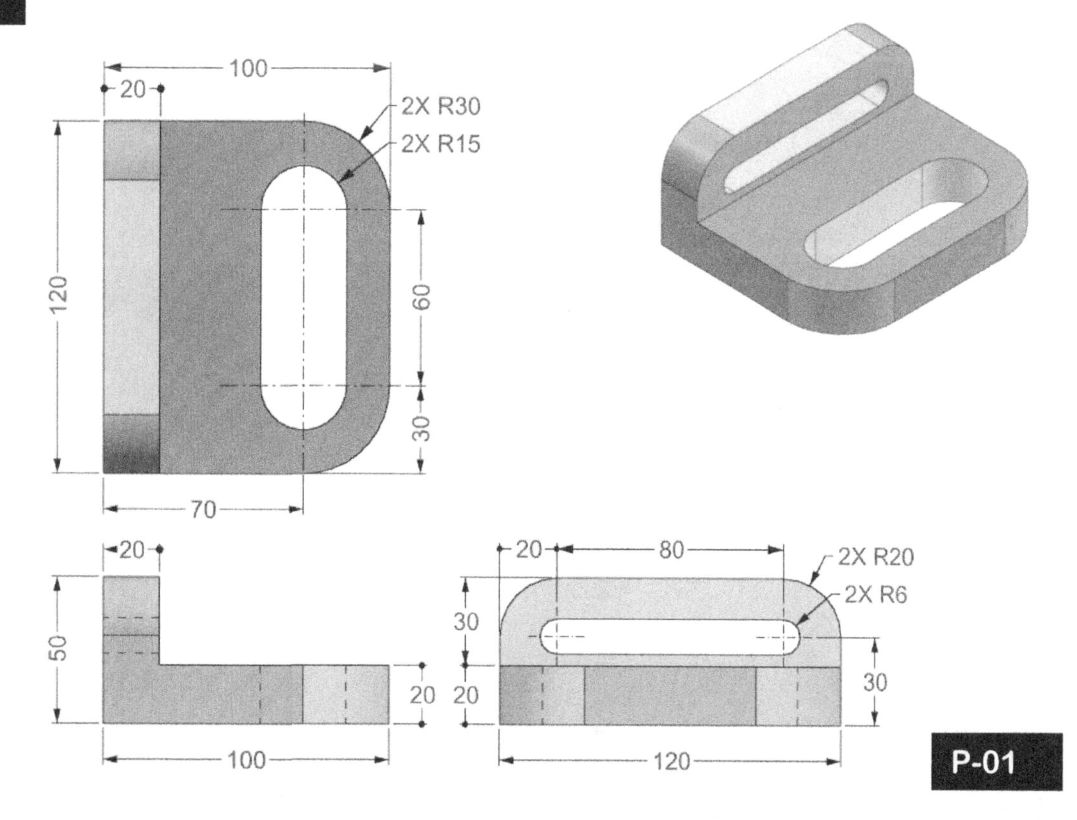

100
20
2X R30
2X R15
120
60
30
70
20
50
20
100
20
80
2X R20
2X R6
30
20
30
120

EX-03

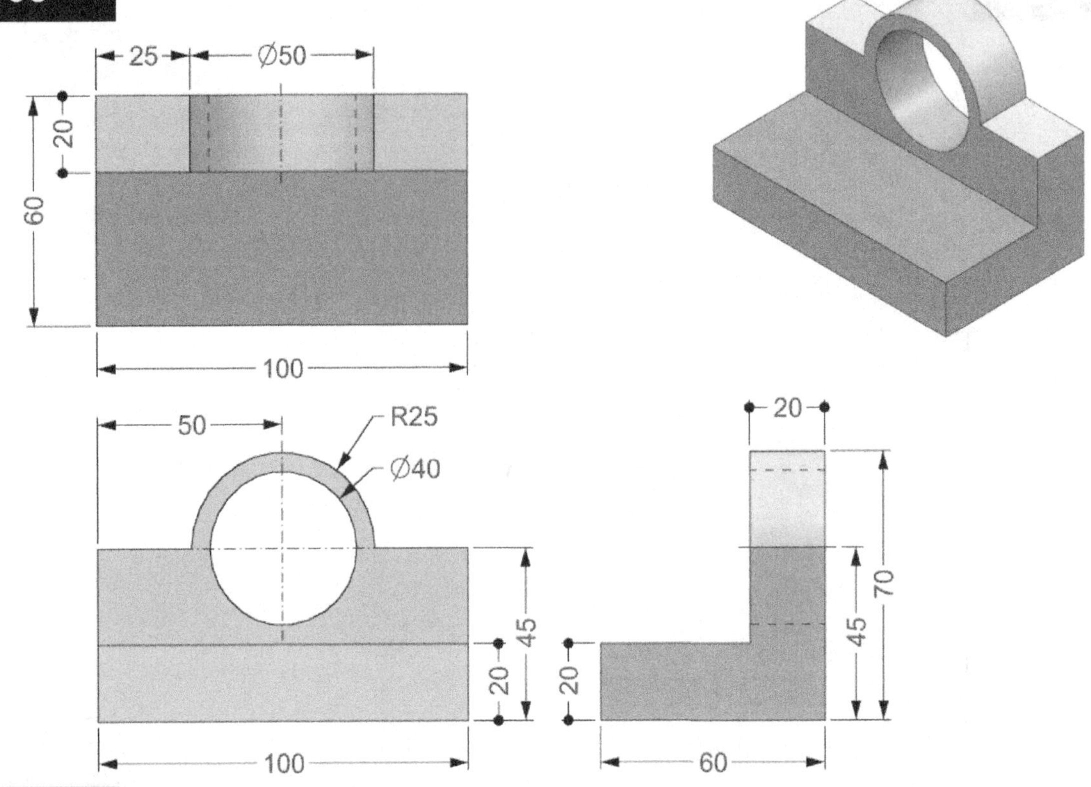

EX-04`

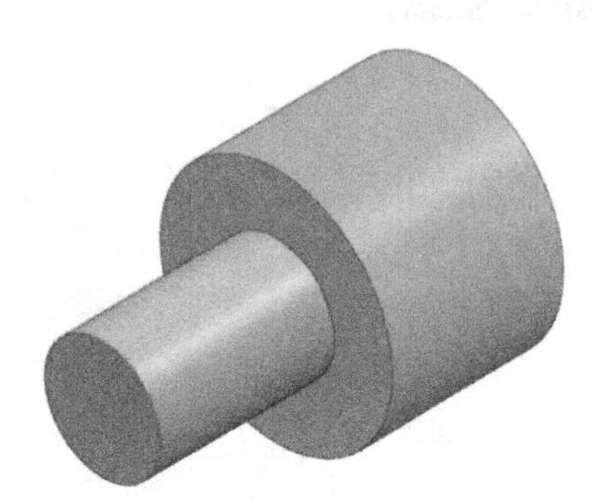

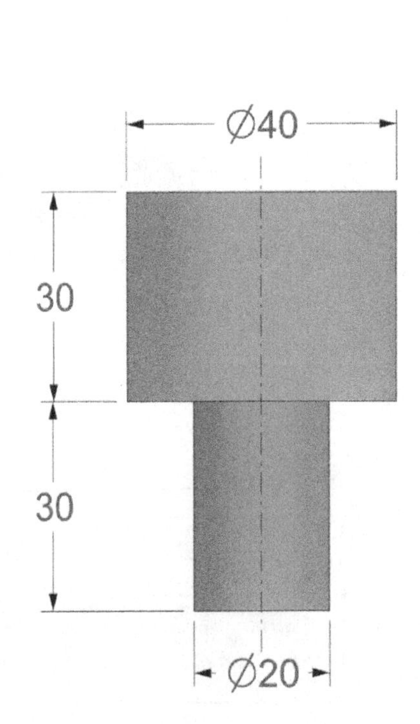

EX-05

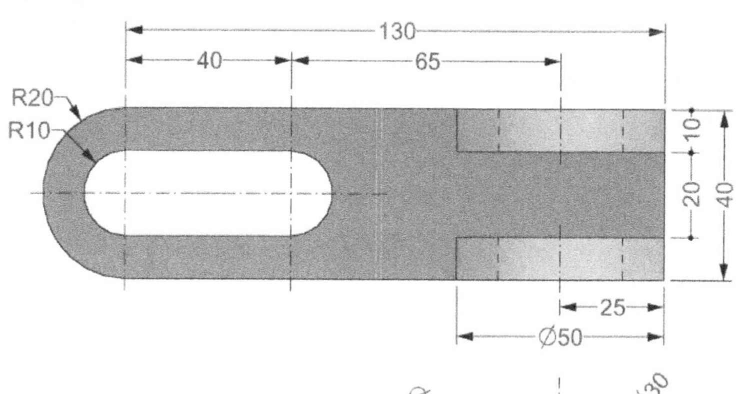

130
40
65
R20
R10
10
20
40
25
Ø50

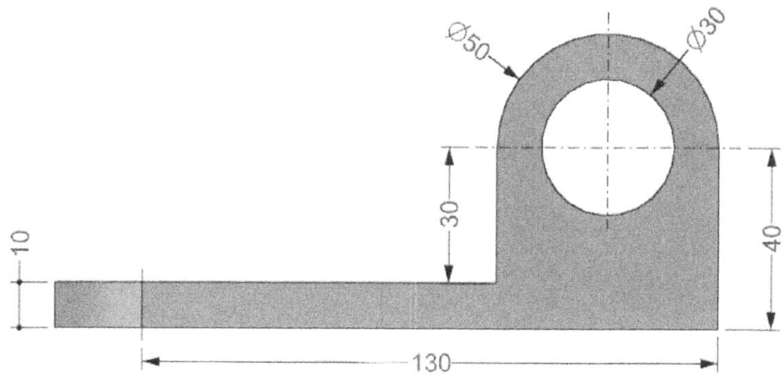

Ø50
Ø30
30
40
10
130

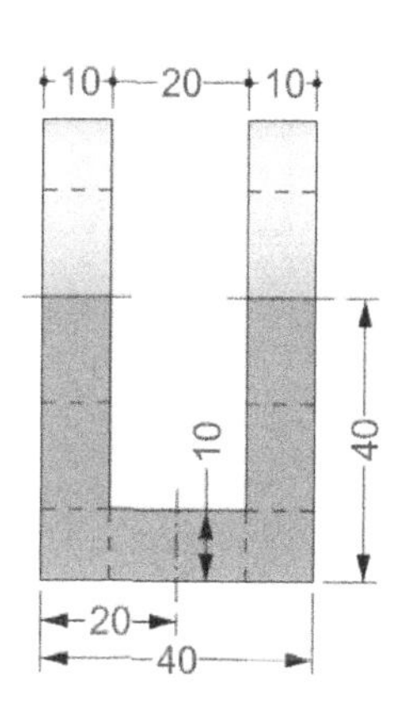

10
20
10
40
10
20
40

EX-06

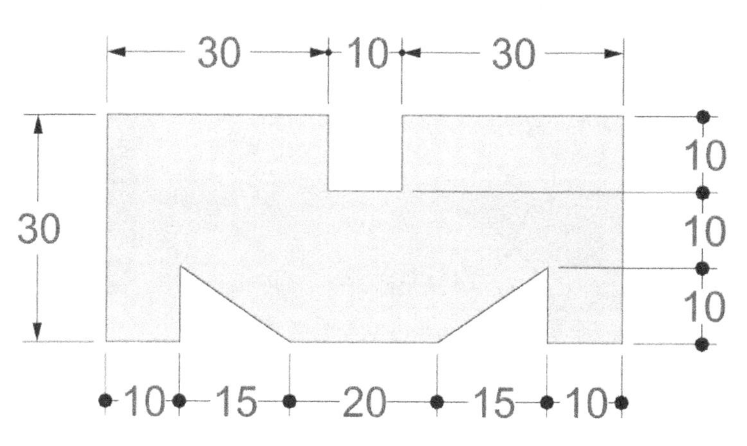

30
10
30
30
10
10
10
10
15
20
15
10

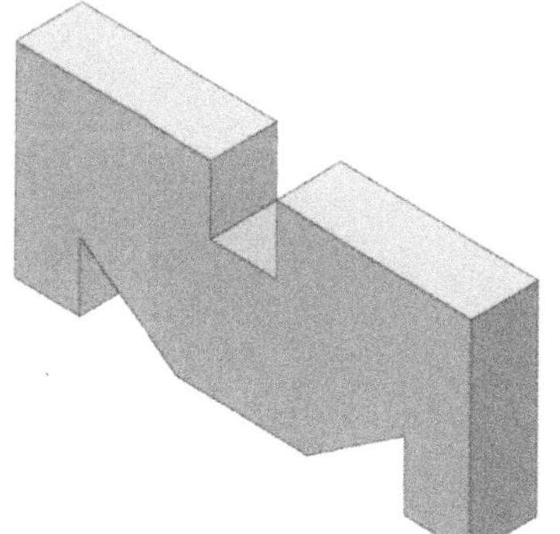

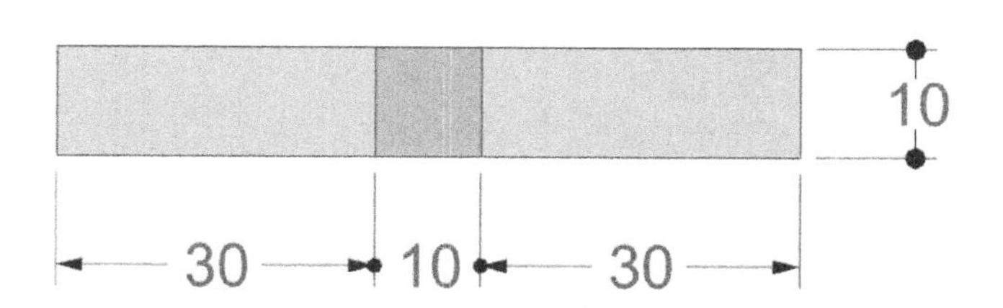

10
30
10
30

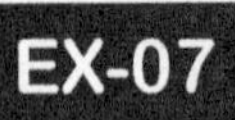
EX-07

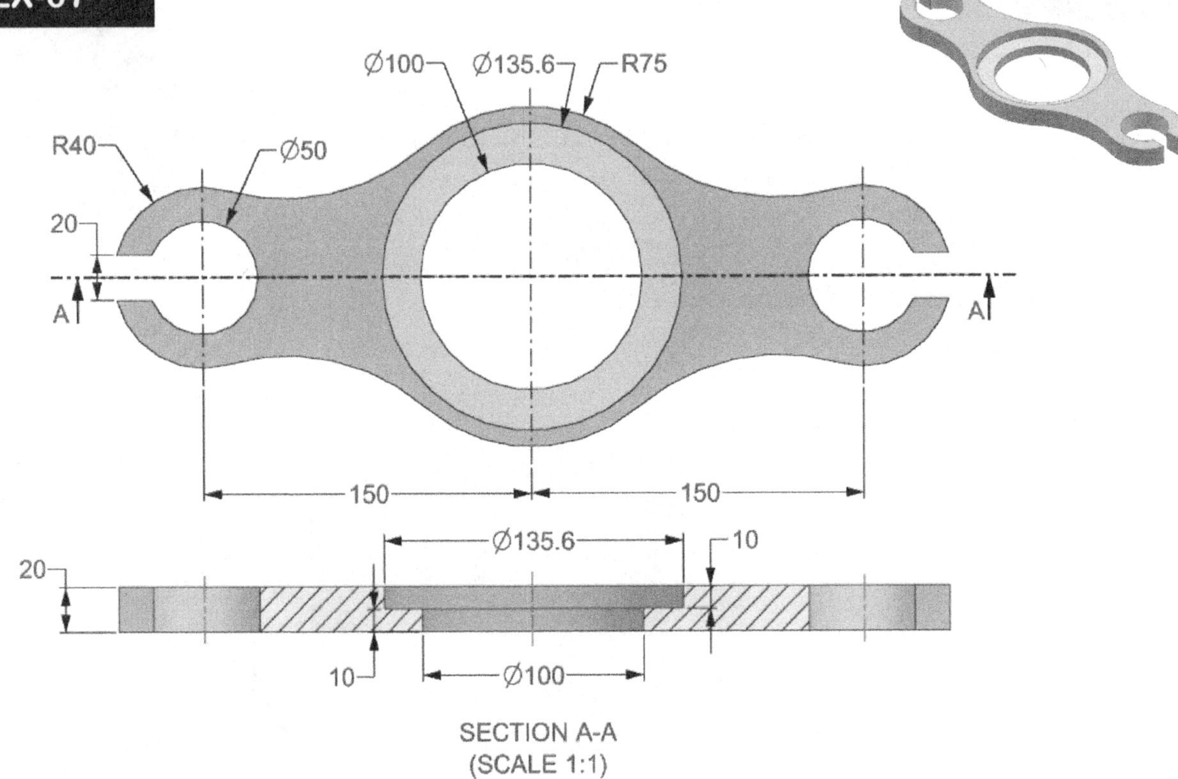
Ø100
Ø135.6
R75
R40
Ø50
20
A
A
150
150
Ø135.6
10
20
10
Ø100
SECTION A-A
(SCALE 1:1)

EX-08

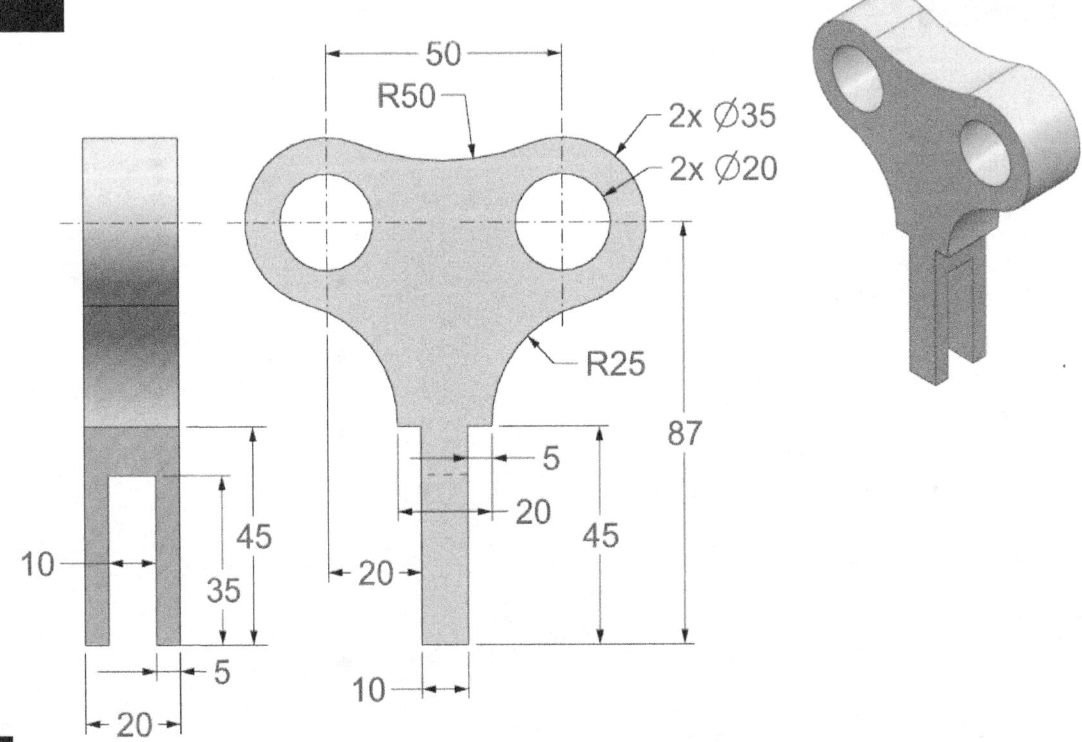
50
R50
2x Ø35
2x Ø20
R25
87
5
20
45
45
35
10
20
5
10
20

EX-09

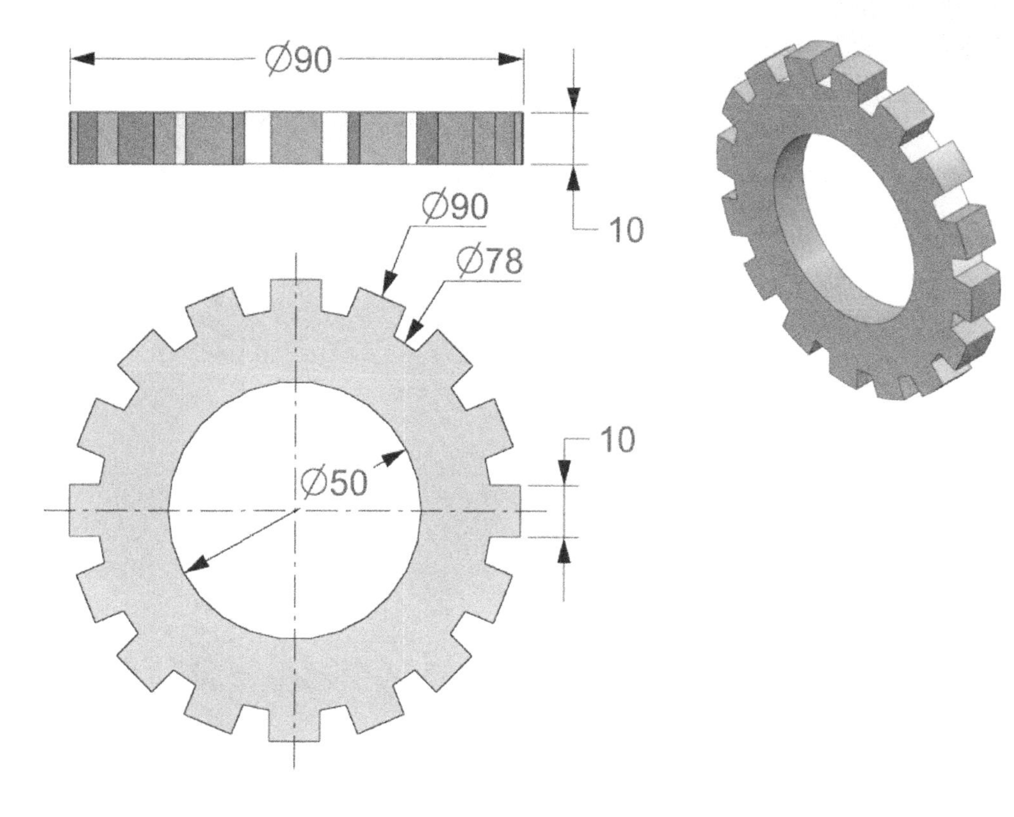

Ø90
10
Ø90
Ø78
10
Ø50

EX-10

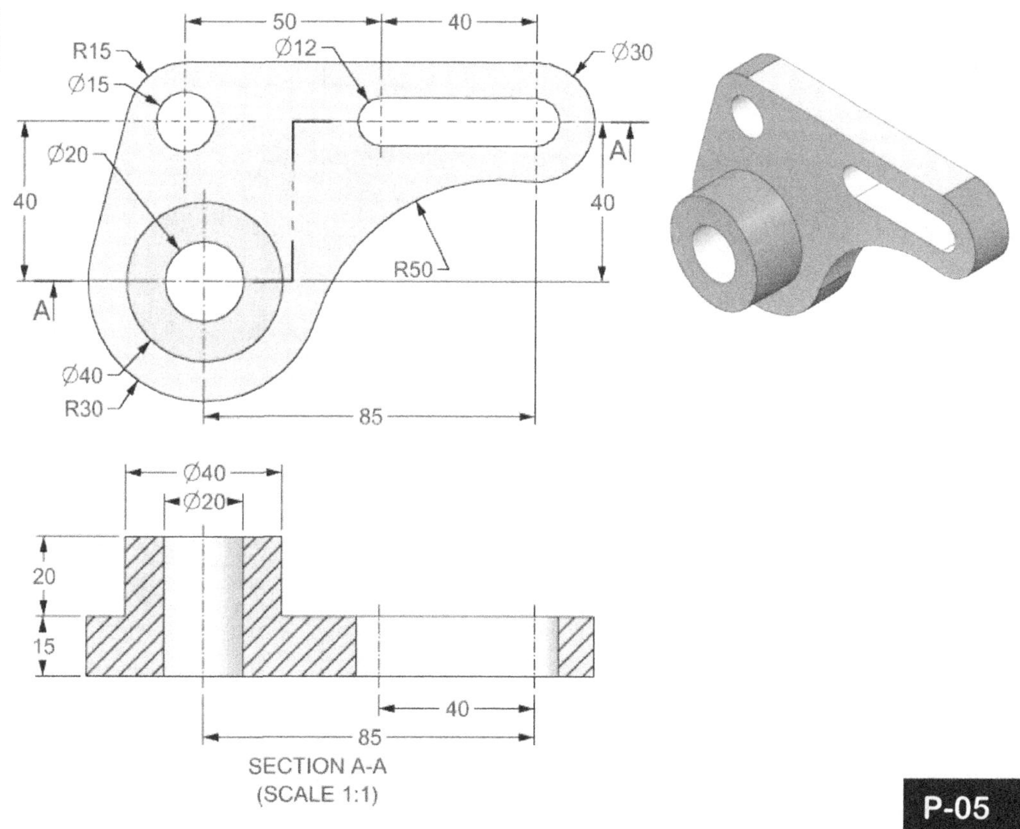

50
40
R15
Ø12
Ø30
Ø15
Ø20
A
40
40
R50
A
Ø40
R30
85
Ø40
Ø20
20
15
40
85
SECTION A-A
(SCALE 1:1)

EX-11

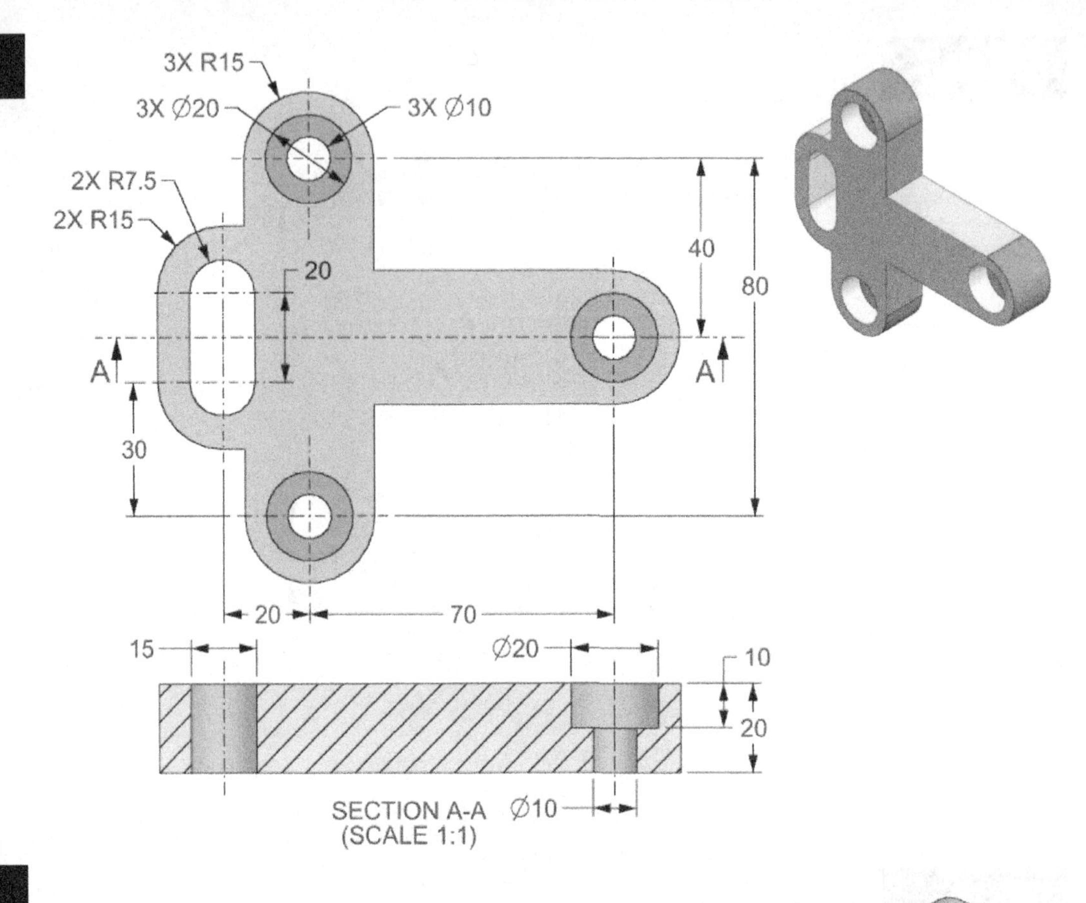
3X R15
3X ∅20
3X ∅10
2X R7.5
2X R15
20
40
80
A
A
30
20
70
15
∅20
10
20
SECTION A-A
(SCALE 1:1)
∅10

EX-12

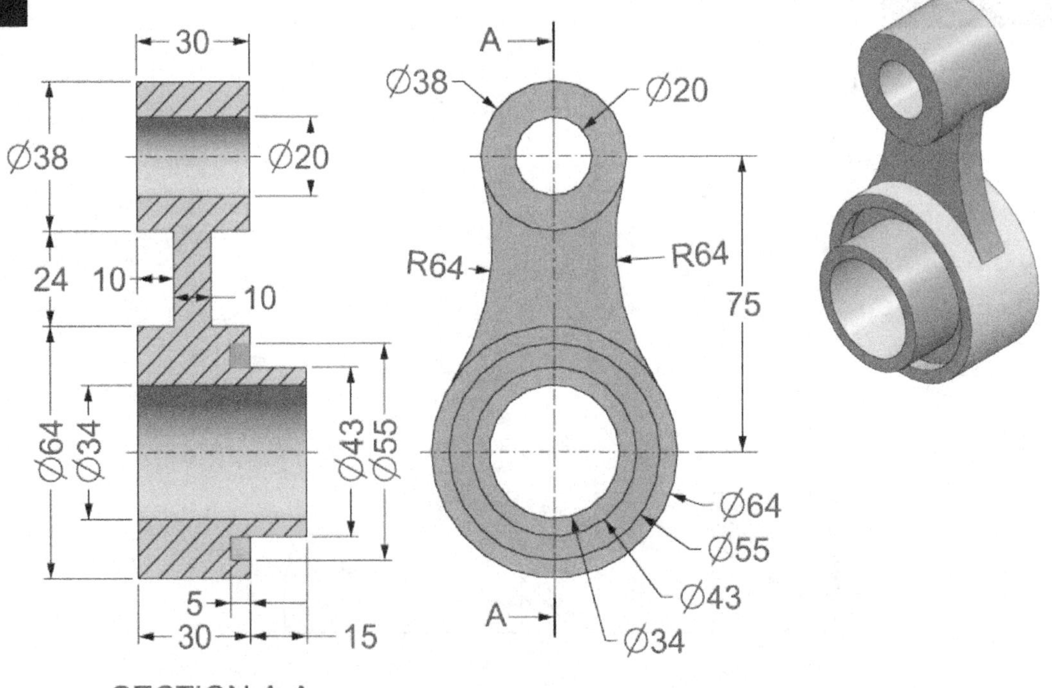
30
∅38
∅20
24
10
10
∅64
∅34
∅43
∅55
5
30
15
SECTION A-A
(SCALE 1:1)
A
∅38
∅20
R64
R64
75
∅64
∅55
∅43
∅34
A

EX-13

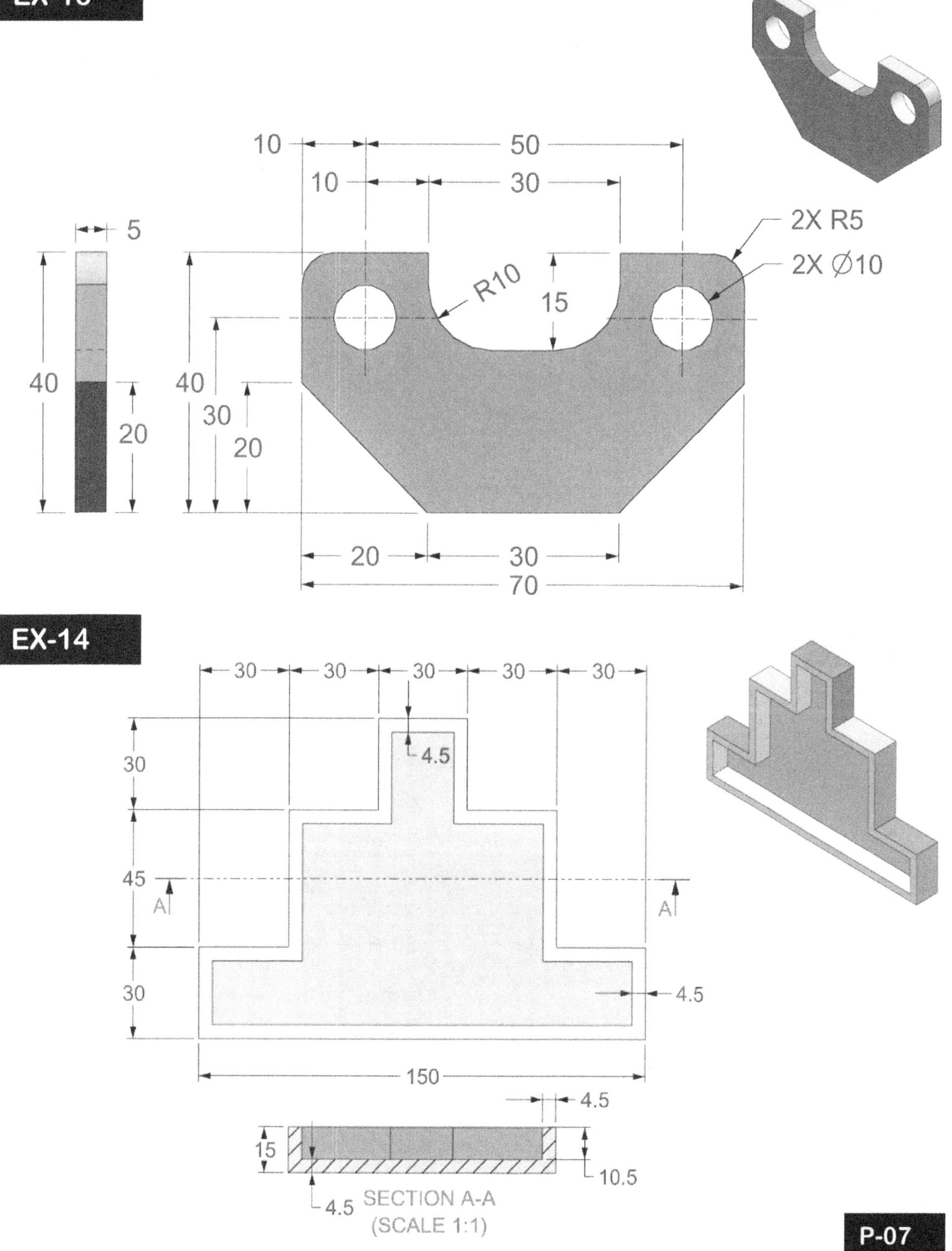

10
50
10
30
5
2X R5
2X ∅10
R10
15
40
20
40
30
20
20
30
70
EX-14
30
30
30
30
30
30
4.5
45
A
A
30
4.5
150
4.5
15
10.5
4.5
SECTION A-A
(SCALE 1:1)

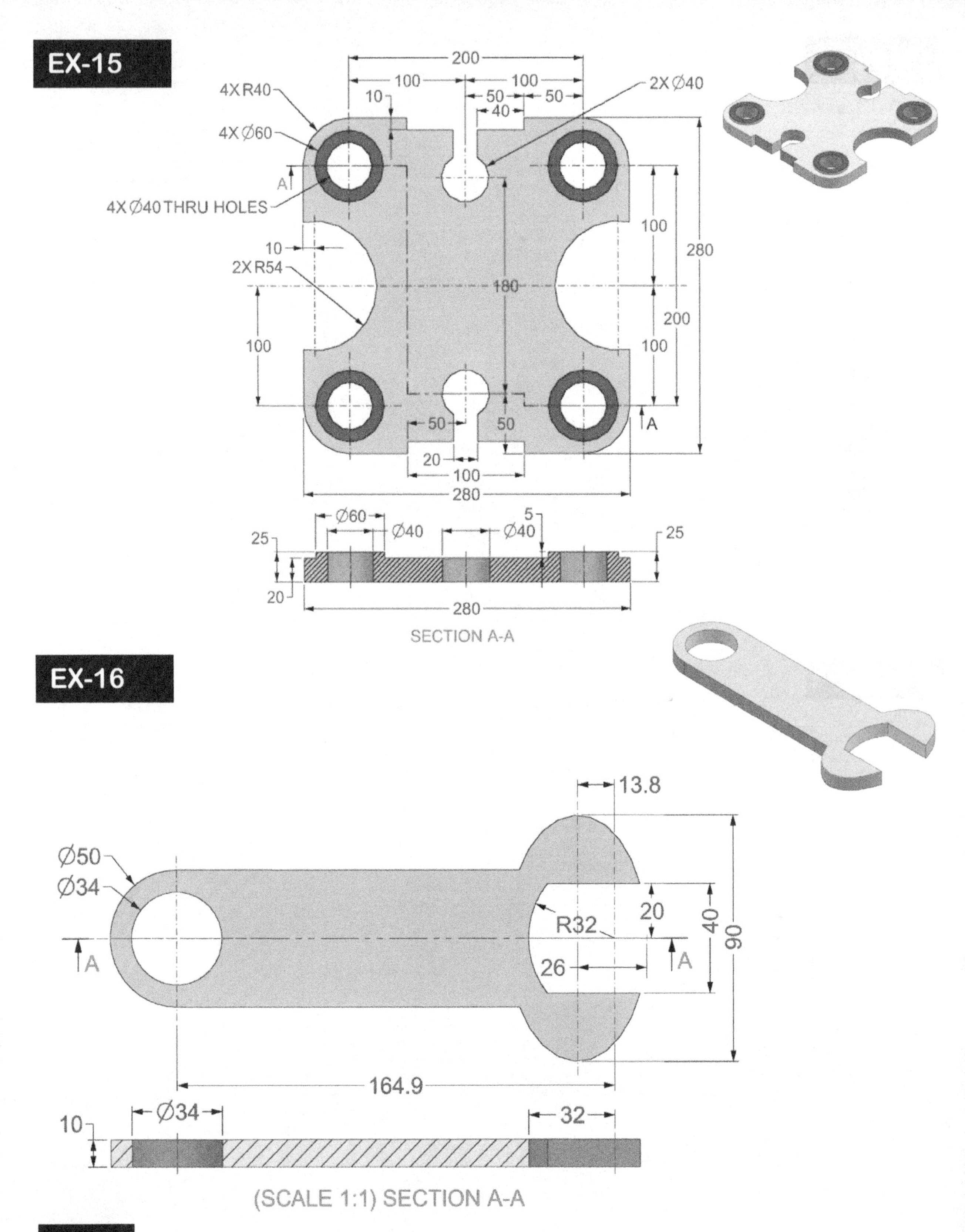
EX-15
200
100
100
10
50
50
40
4X R40
4X Ø60
2X Ø40
A
4X Ø40 THRU HOLES
10
2X R54
180
100
100
100
200
280
50
50
A
20
100
280
Ø60
Ø40
Ø40
5
25
25
20
280
SECTION A-A
EX-16
13.8
Ø50
Ø34
R32
20
40
90
A
26
A
164.9
Ø34
32
10
(SCALE 1:1) SECTION A-A

EX-17

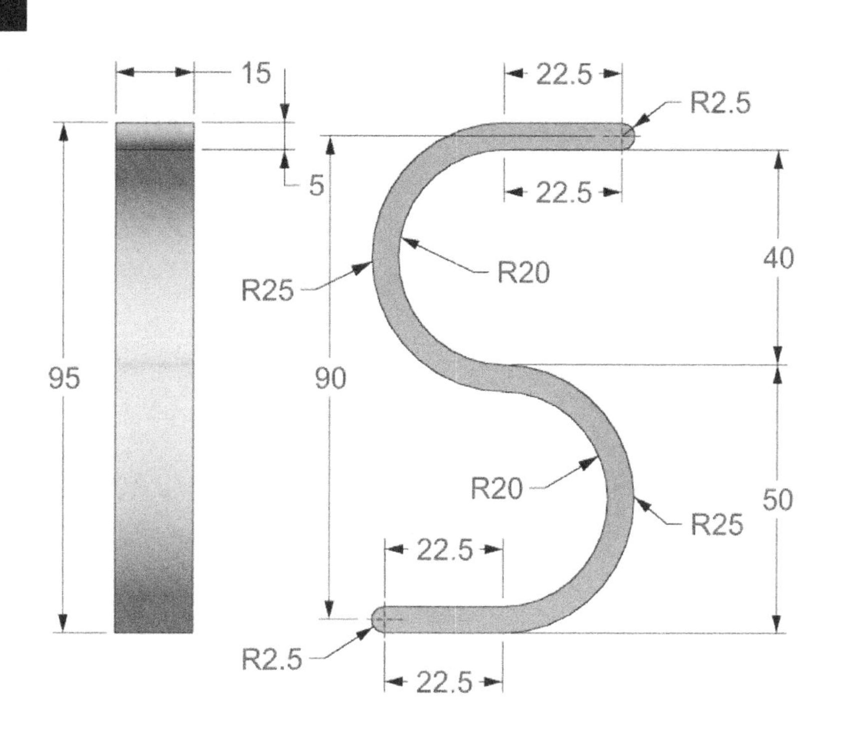

15
5
95
22.5
R2.5
22.5
R20
R25
40
90
R20
R25
50
22.5
R2.5
22.5

EX-18

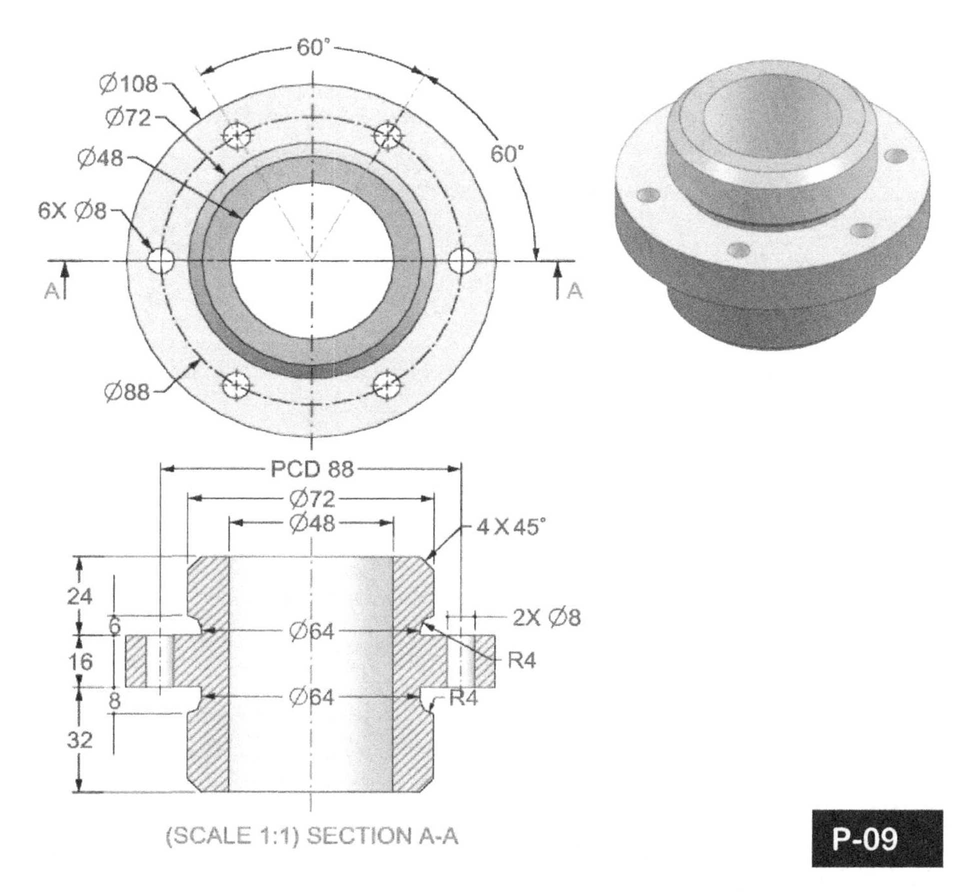

60°
Ø108
Ø72
Ø48
60°
6X Ø8
A
A
Ø88
PCD 88
Ø72
Ø48
4 X 45°
24
6
2X Ø8
Ø64
16
R4
8
Ø64
R4
32
(SCALE 1:1) SECTION A-A

EX-19

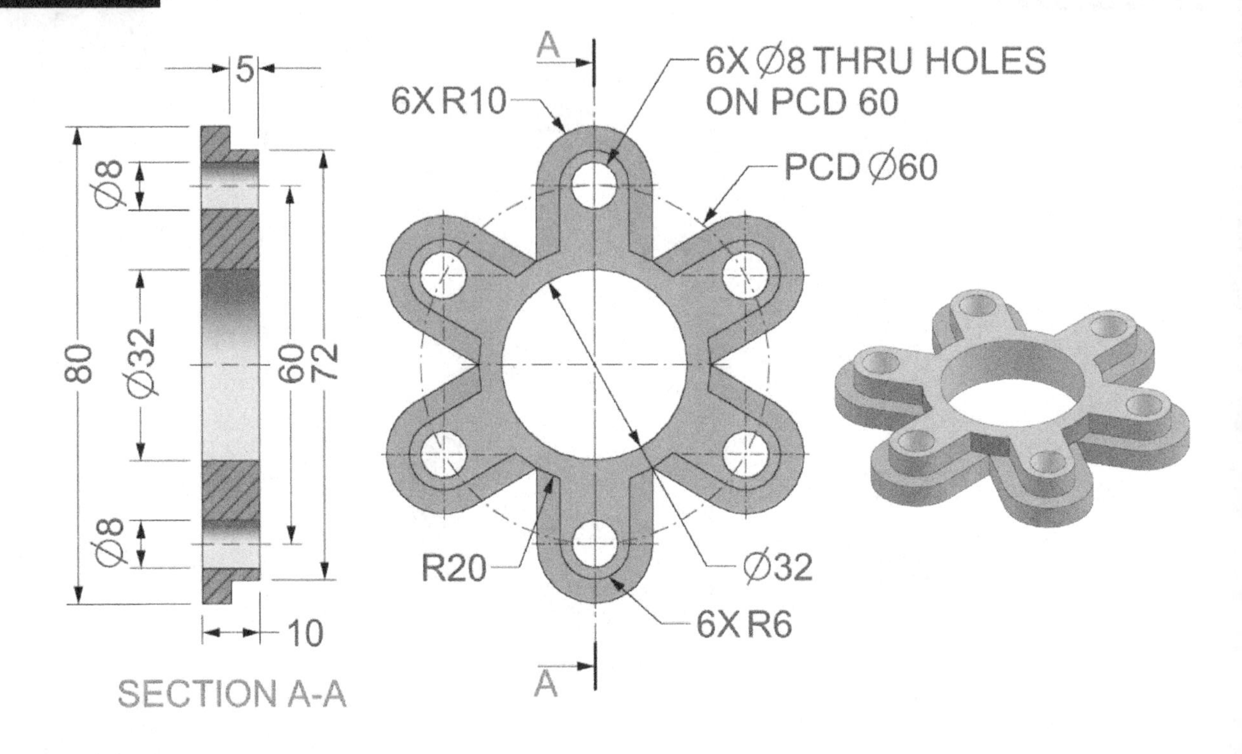
5
6X R10
A
6X Ø8 THRU HOLES
ON PCD 60
PCD Ø60
Ø8
80
Ø32
60
72
Ø8
10
R20
Ø32
6X R6
A
SECTION A-A

EX-20

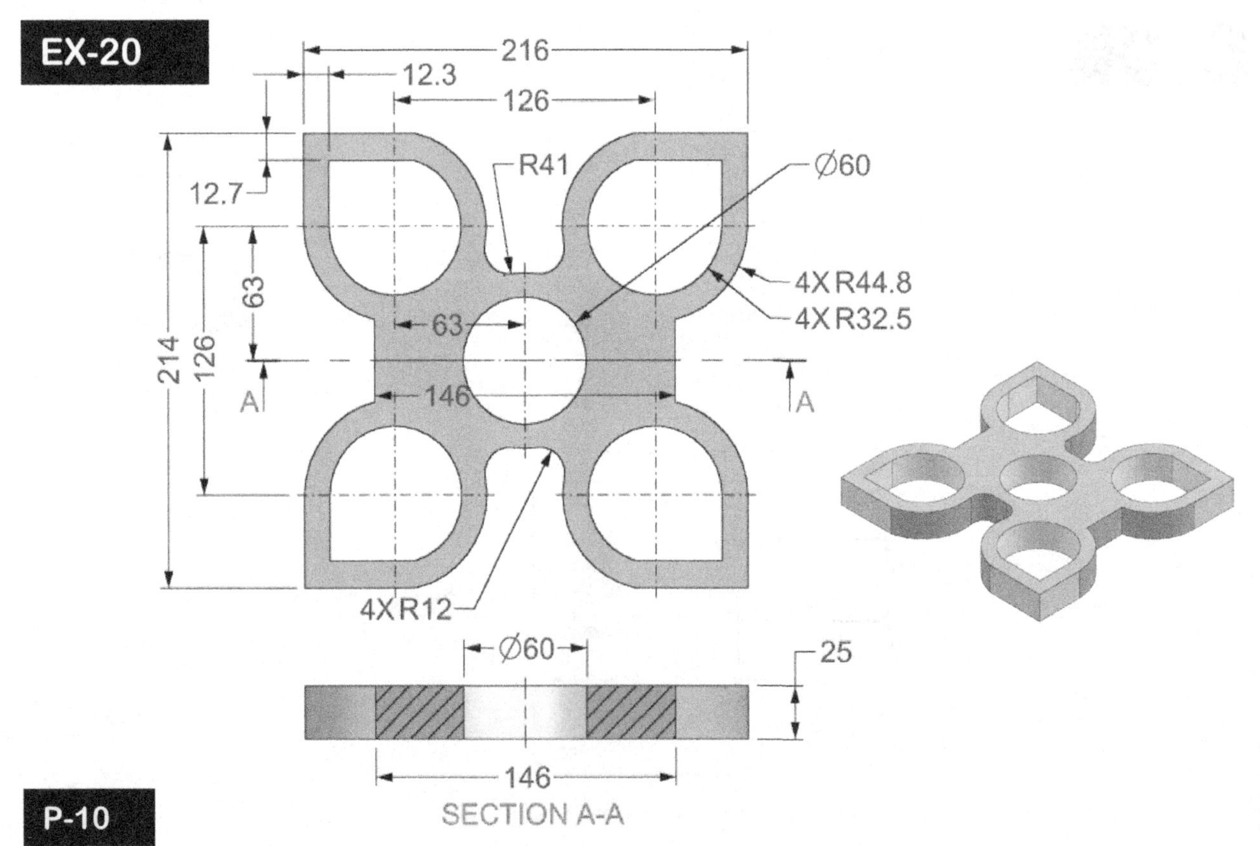
216
12.3
126
12.7
R41
Ø60
4X R44.8
4X R32.5
63
214
126
63
A
A
146
4X R12
Ø60
25
146
SECTION A-A

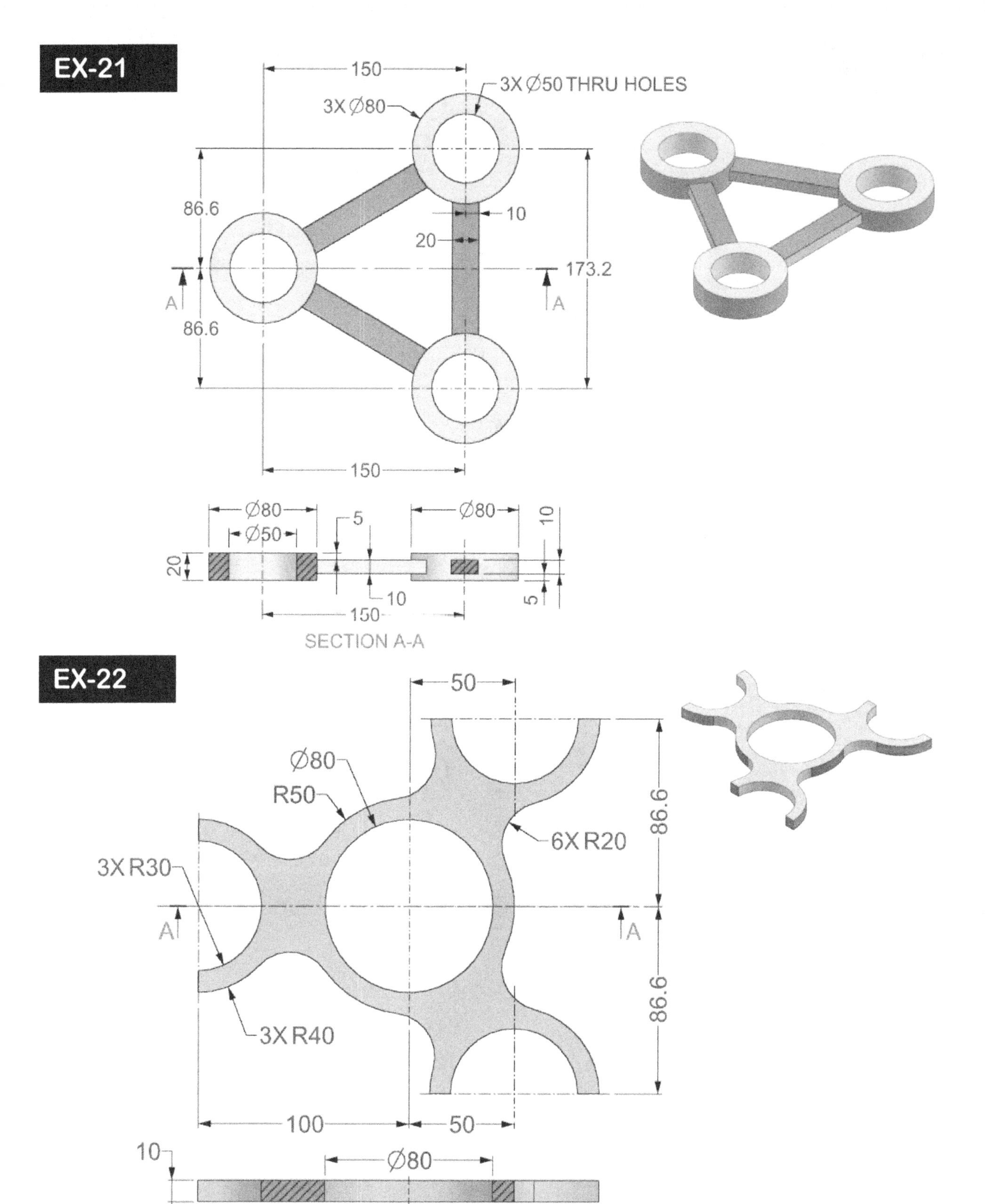

EX-21
150
3X Ø50 THRU HOLES
3X Ø80
86.6
10
20
173.2
A
A
86.6
150
Ø80
Ø50
5
Ø80
10
20
10
150
5
SECTION A-A
EX-22
50
Ø80
R50
86.6
6X R20
3X R30
A
A
86.6
3X R40
100
50
10
Ø80
SECTION A-A

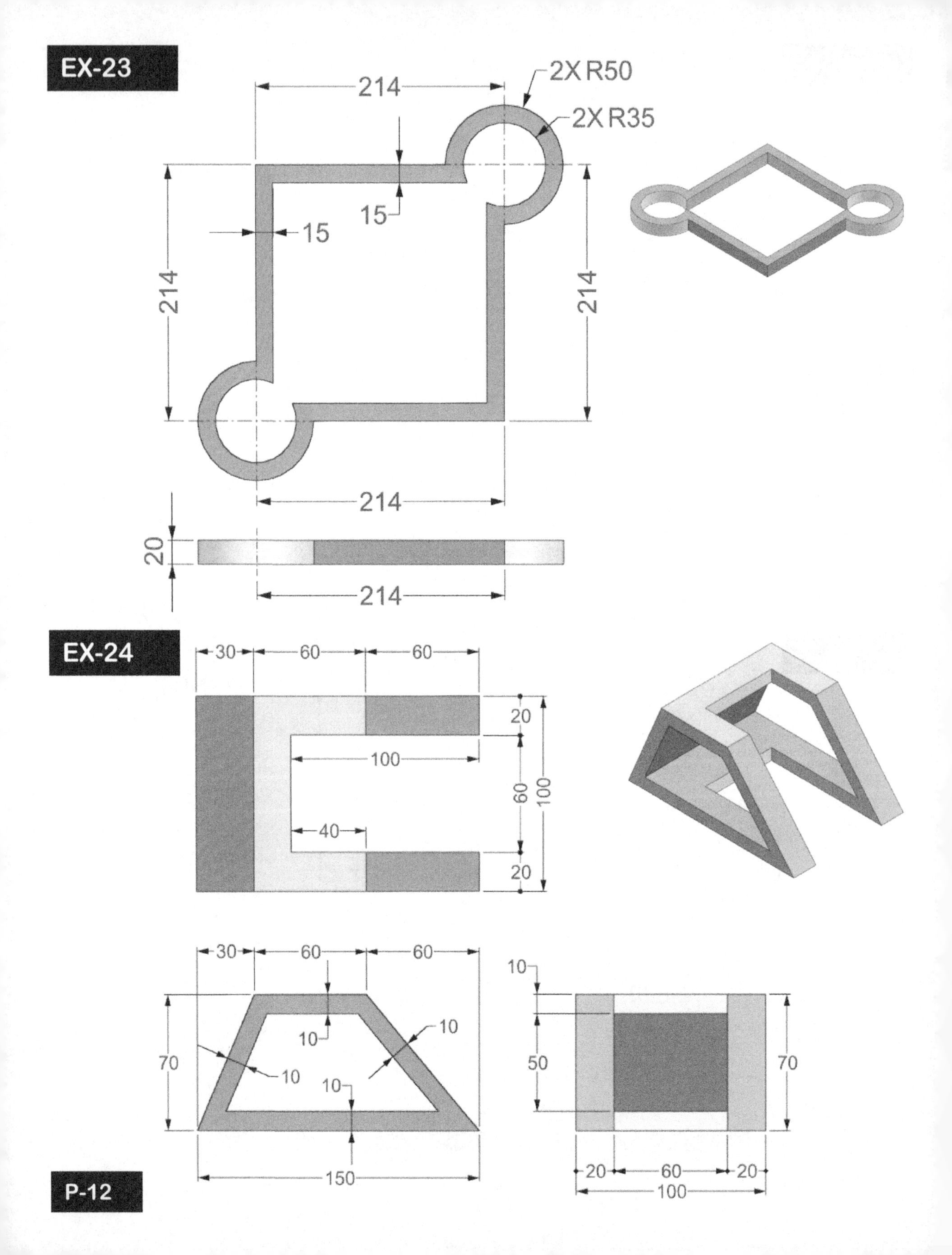

EX-23
2X R50
2X R35
214
15
15
214
214
214
20
214
EX-24
30
60
60
20
100
60
100
40
20
30
60
60
70
10
10
10
10
150
10
50
70
20
60
20
100
P-12

EX-25

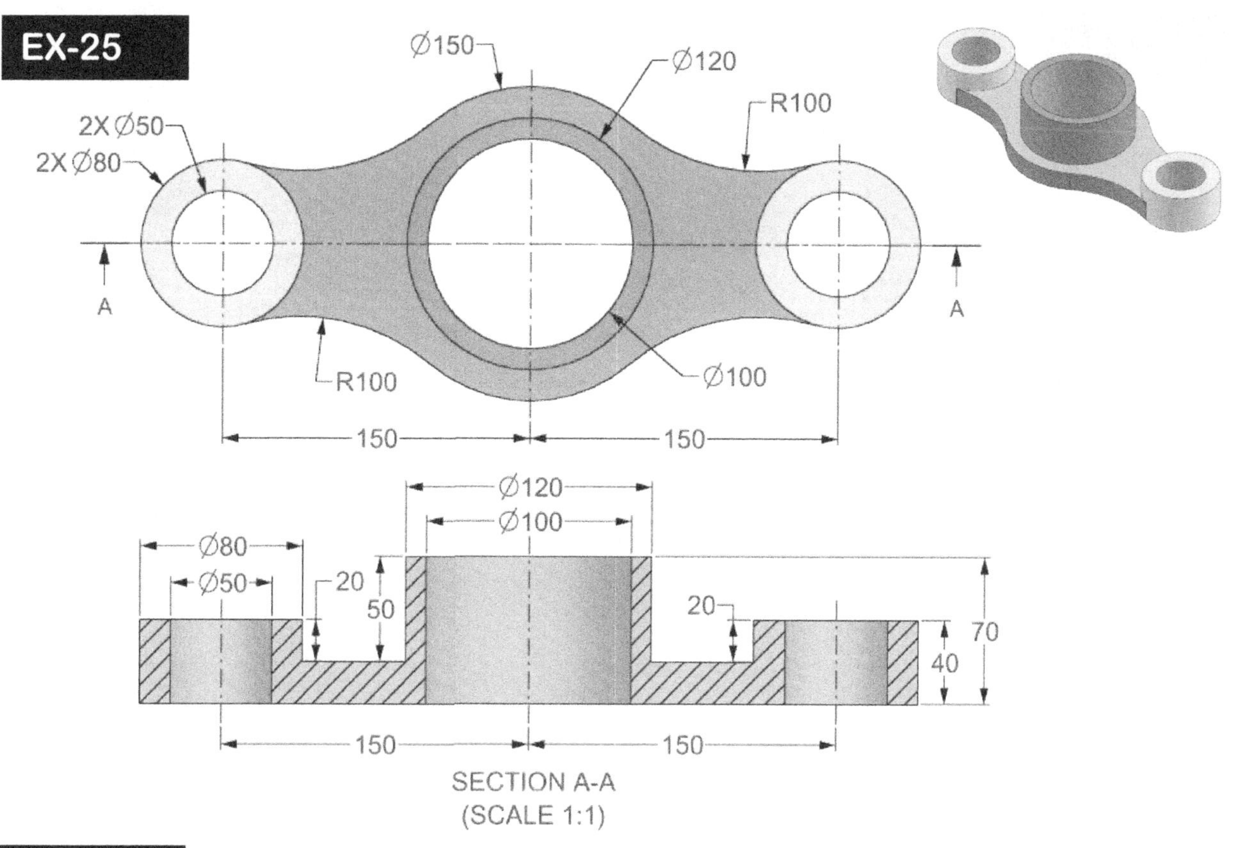

Ø150
Ø120
R100
2X Ø50
2X Ø80
A
A
R100
Ø100
150
150
Ø120
Ø100
Ø80
Ø50
20
50
20
70
40
150
150
SECTION A-A
(SCALE 1:1)

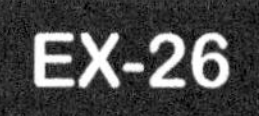

EX-26

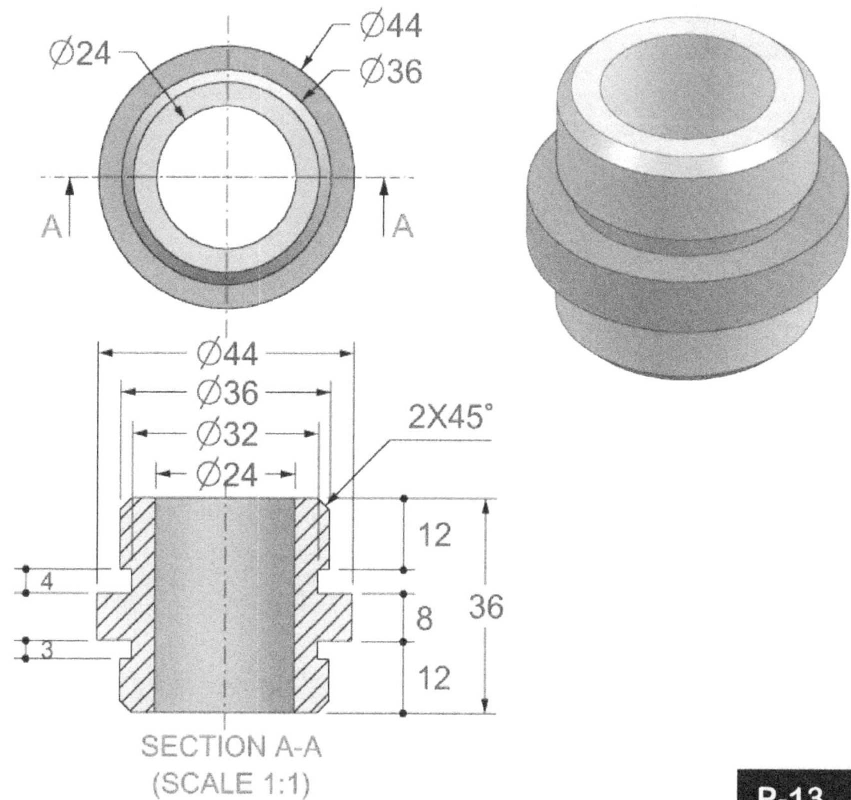

Ø44
Ø24
Ø36
A
A
Ø44
Ø36
Ø32
Ø24
2X45°
12
4
8
36
3
12
SECTION A-A
(SCALE 1:1)

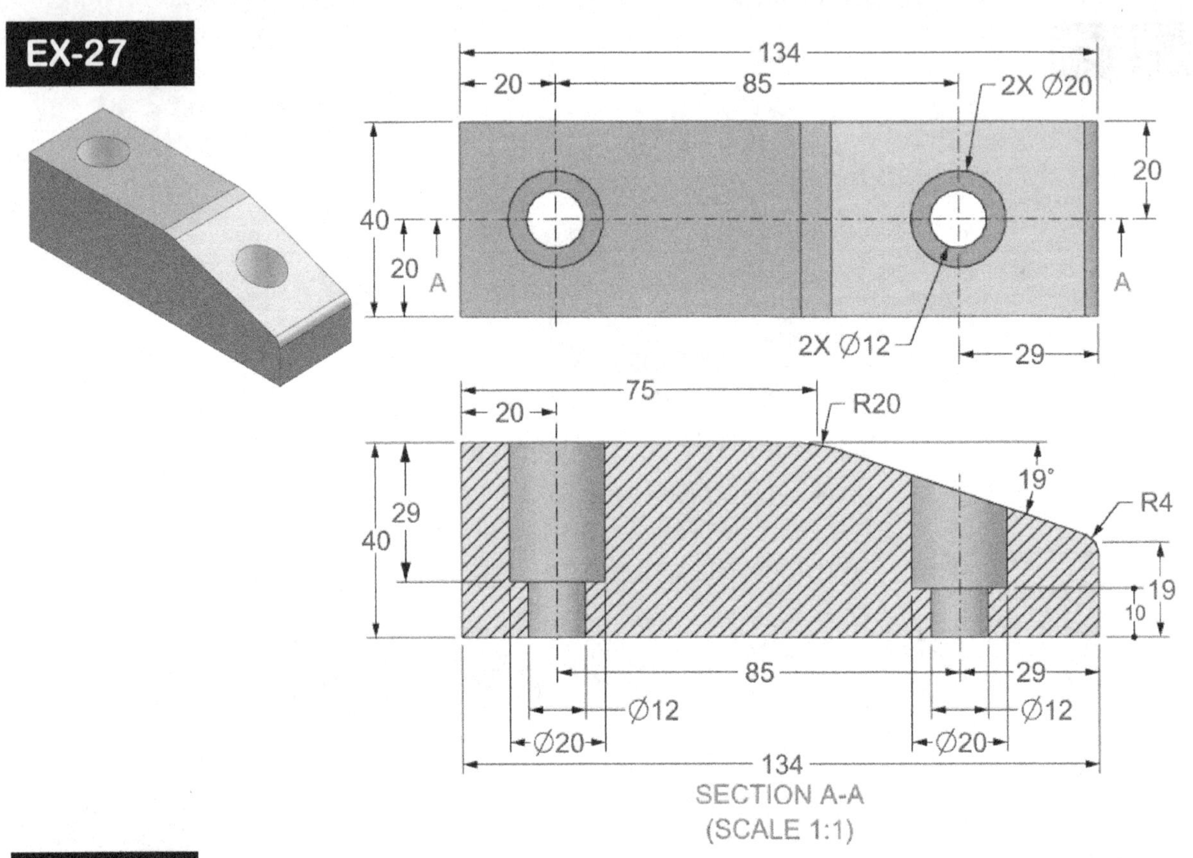
EX-27
134
20
85
2X Ø20
20
40
20
A
A
2X Ø12
29
75
20
R20
19°
R4
29
40
19
10
85
29
Ø12
Ø12
Ø20
Ø20
134
SECTION A-A
(SCALE 1:1)

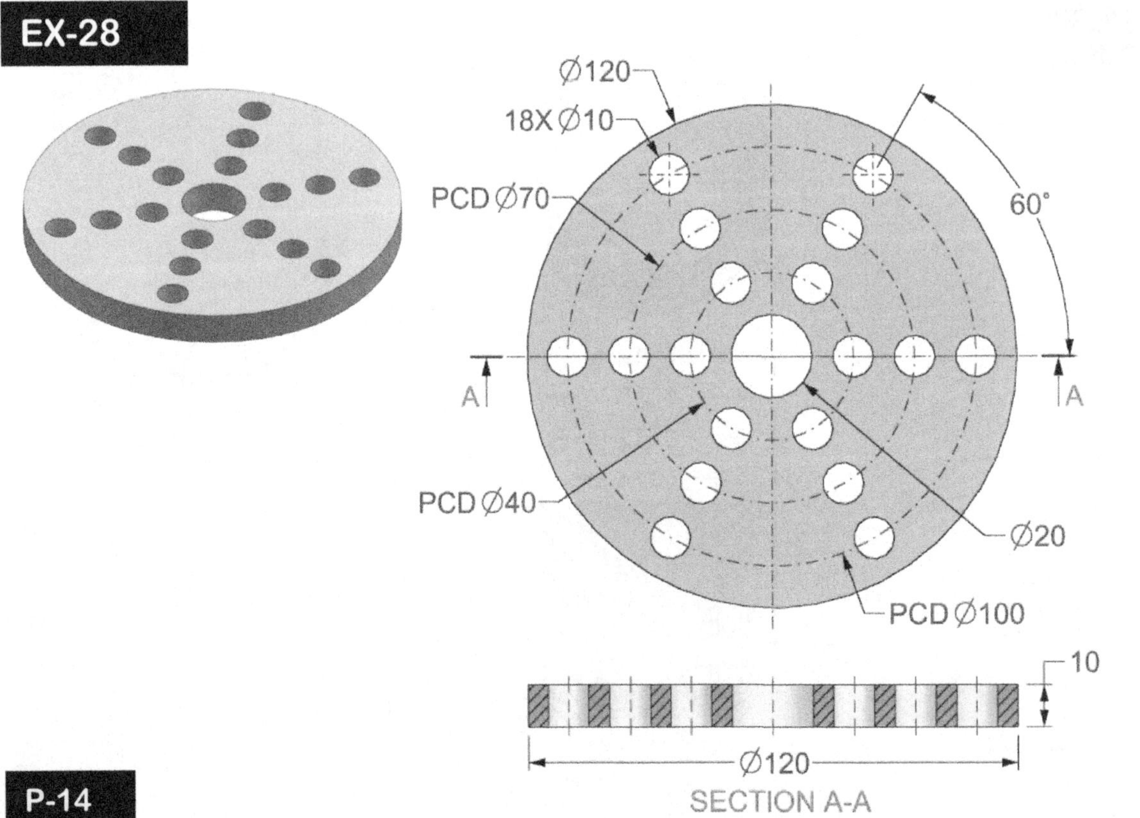
EX-28
Ø120
18X Ø10
PCD Ø70
60°
A
A
PCD Ø40
Ø20
PCD Ø100
10
Ø120
SECTION A-A

EX-29

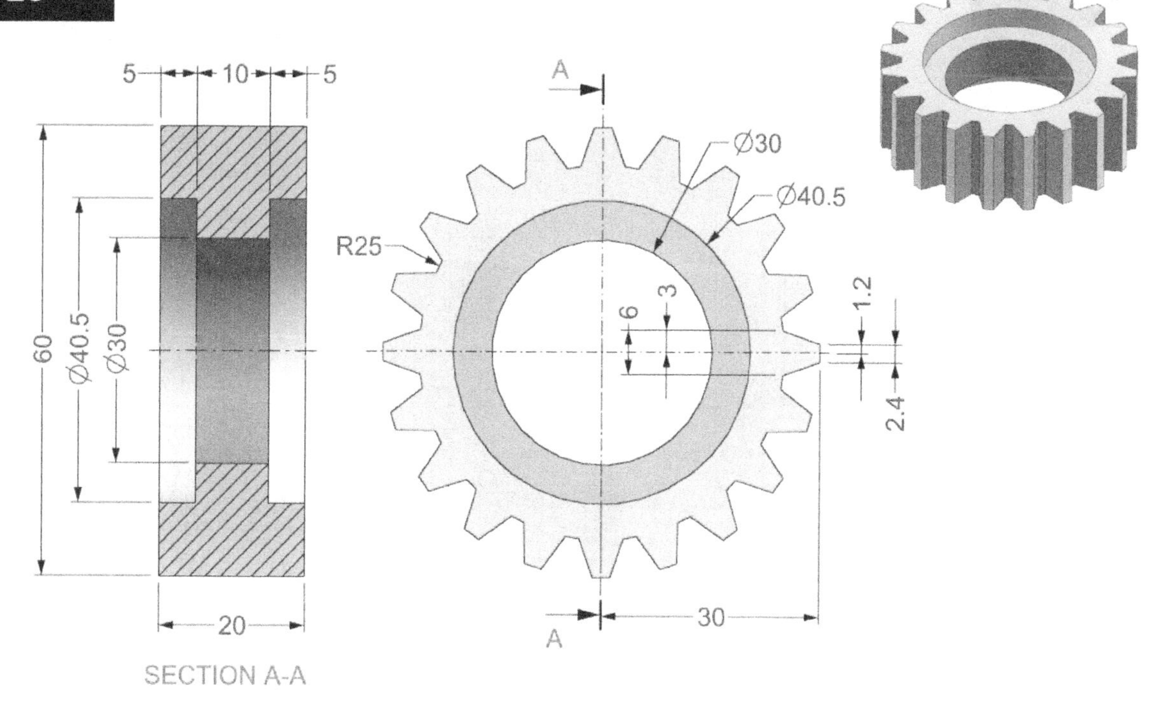

5
10
5
60
Ø40.5
Ø30
20
SECTION A-A
A
Ø30
Ø40.5
R25
6
3
1.2
2.4
30
A

EX-30

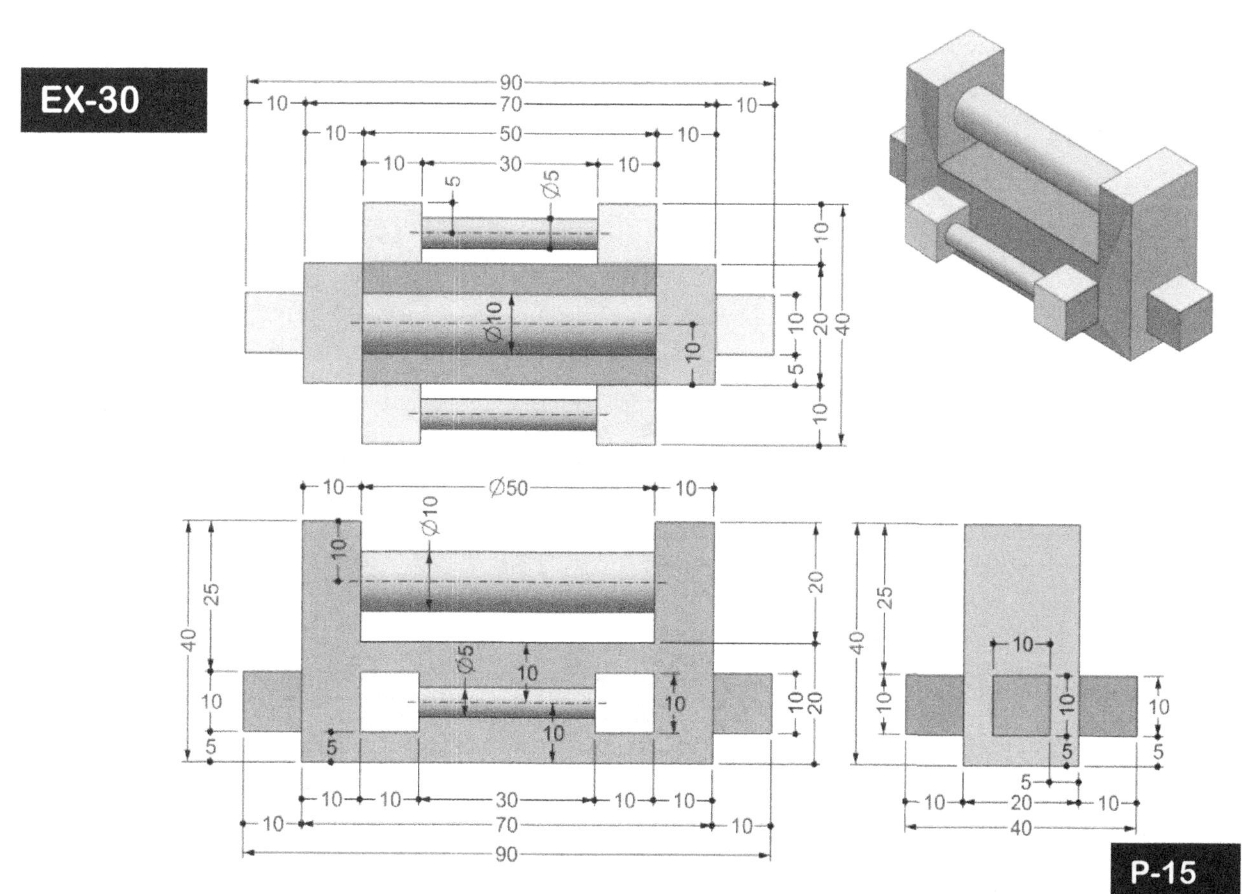

90
10
70
10
10
50
10
10
30
10
5
Ø5
Ø10
10
20
40
Ø50
Ø10
25
Ø5
5

EX-31

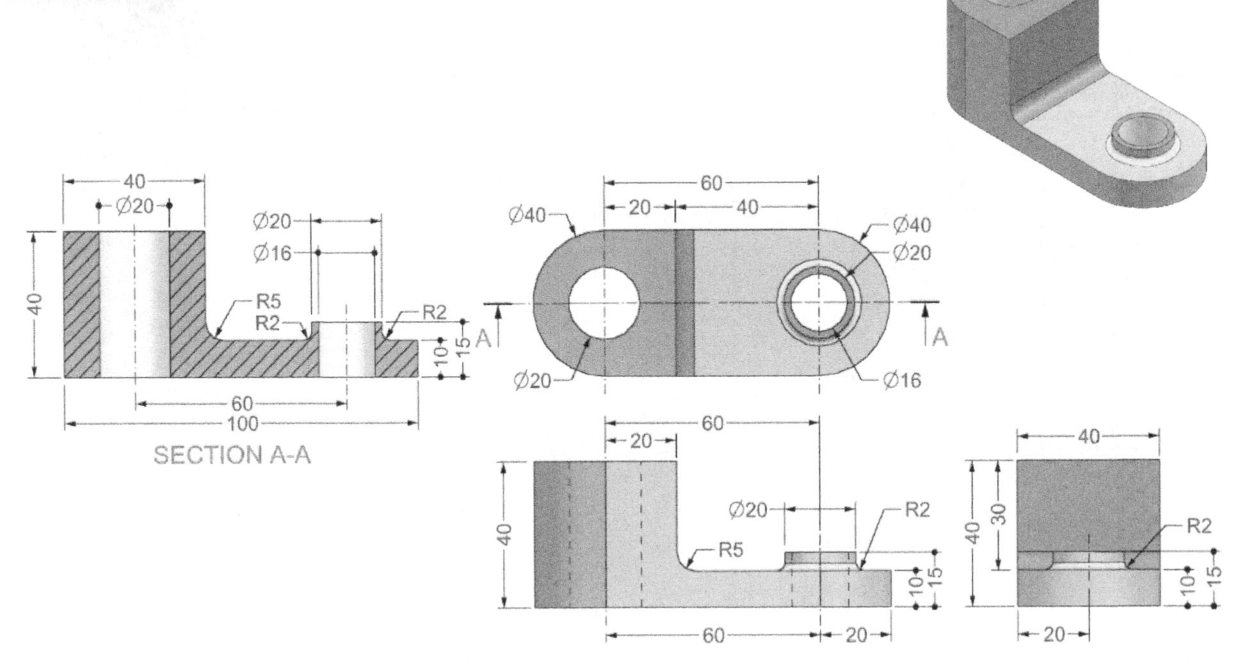
40
Ø20
Ø20
Ø16
R5
R2
R2
40
10
15
A
60
100
SECTION A-A
60
20
40
Ø40
Ø40
Ø20
Ø20
Ø16
A
60
20
40
Ø20
R2
R5
10
15
60
20
40
40
30
R2
10
15
20

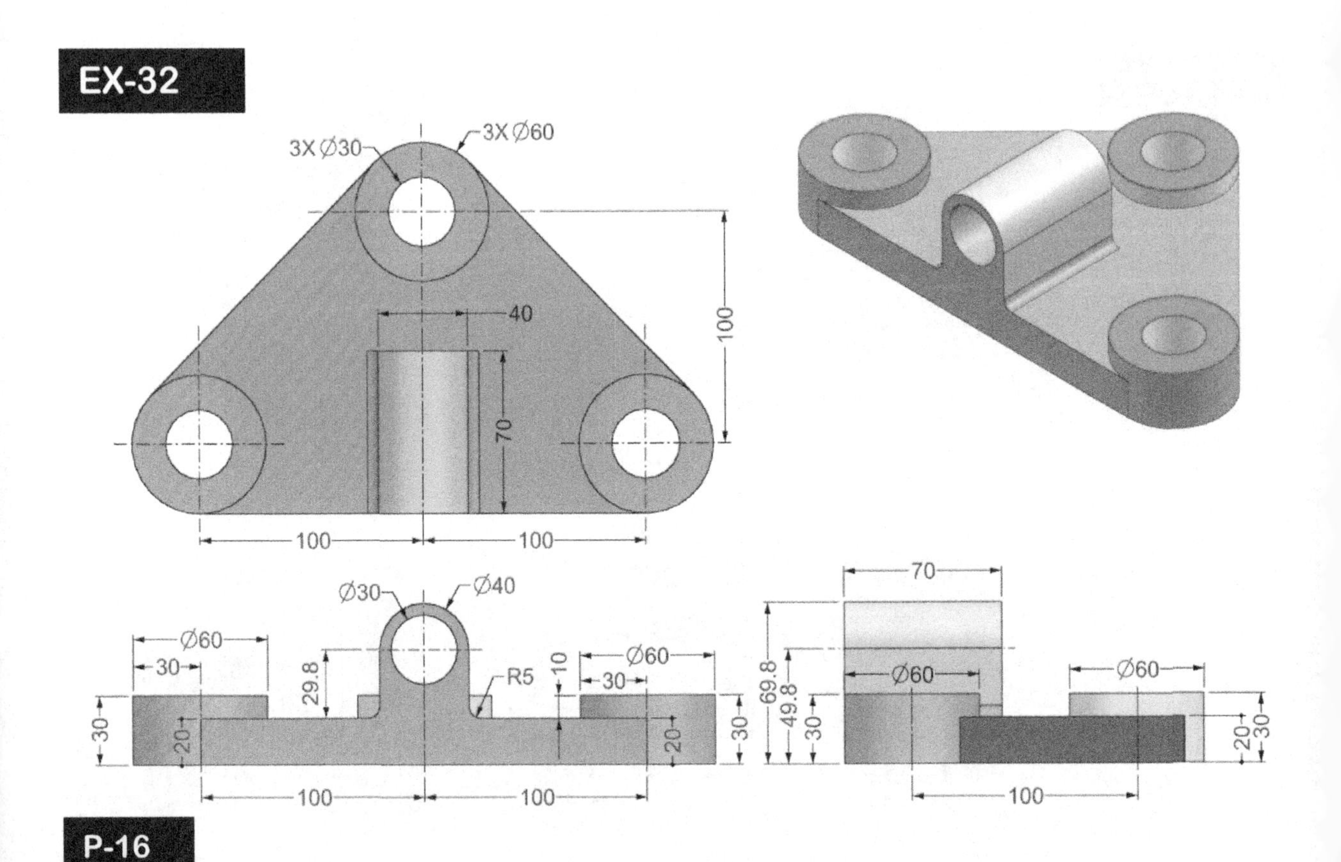
EX-32
3X Ø30
3X Ø60
40
100
70
100
100
Ø30
Ø40
Ø60
30
29.8
R5
10
Ø60
30
30
20
20
30
100
100
70
69.8
49.8
30
Ø60
Ø60
20
30
100

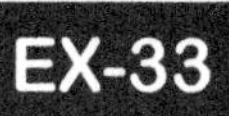
EX-33

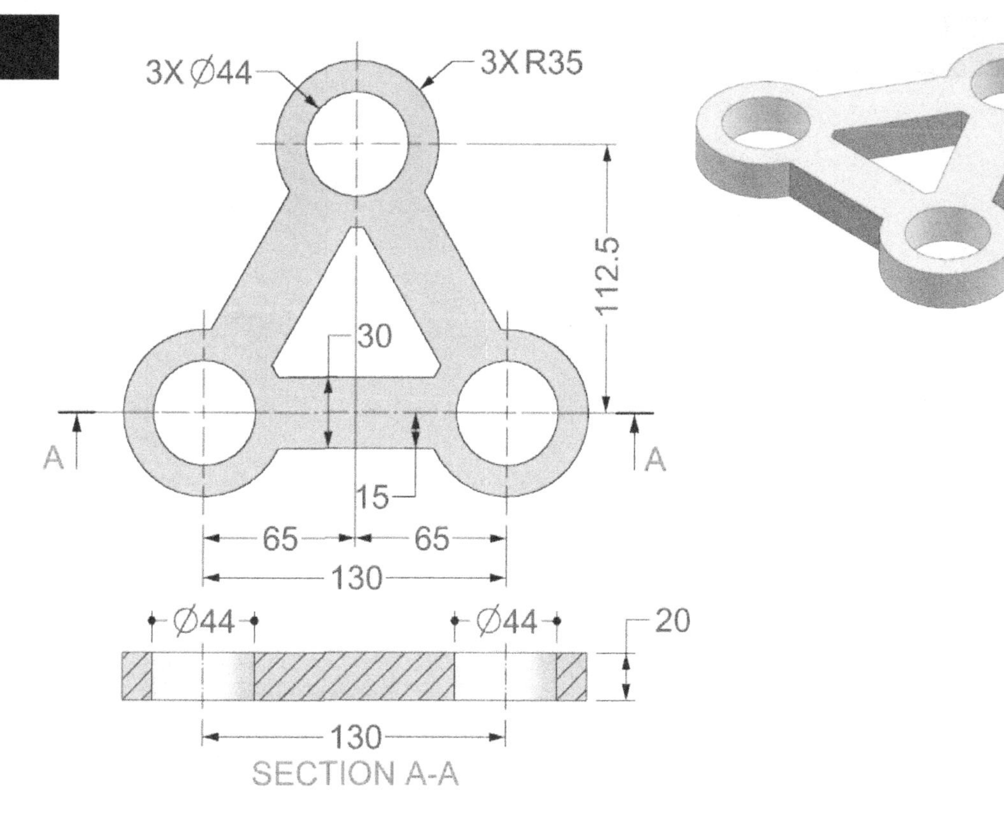
3X Ø44
3X R35
112.5
30
15
A
A
65
65
130
Ø44
Ø44
20
130
SECTION A-A

EX-34

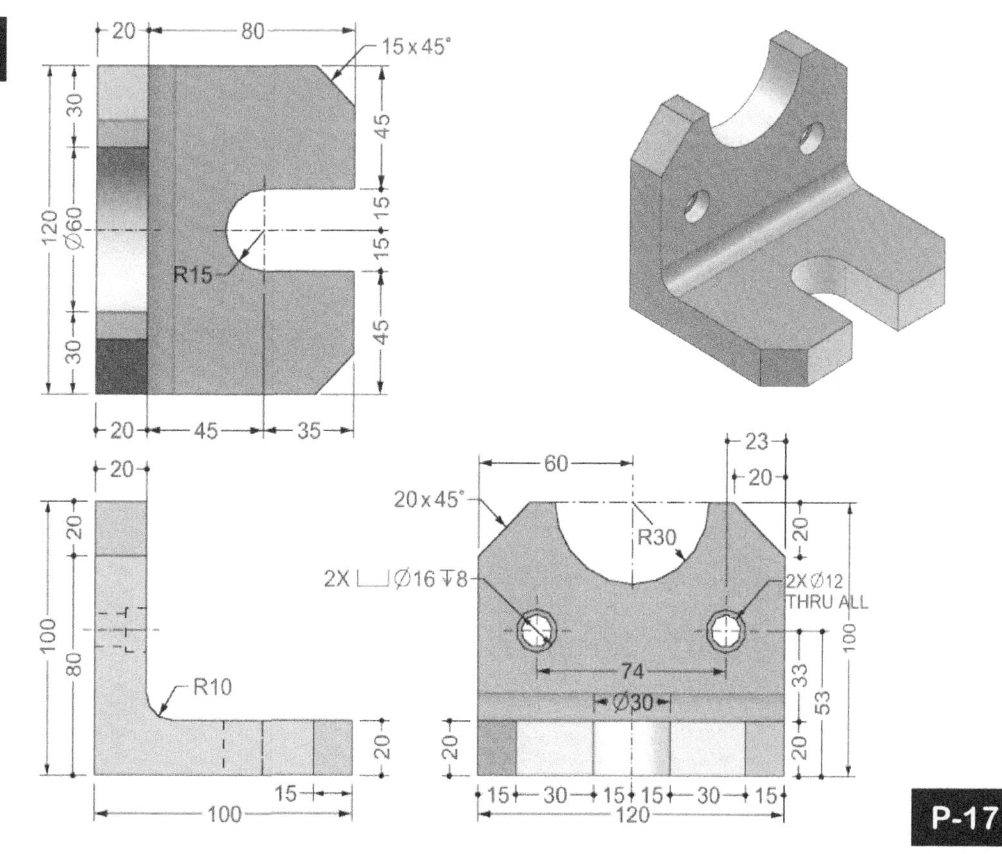
20
80
15 x 45°
30
45
120
Ø60
15
15
R15
45
30
20
45
35
20
20
100
80
R10
20
15
100
60
23
20
20 x 45°
R30
20
2X ⌴ Ø16 ↧8
2X Ø12
THRU ALL
100
74
33
53
Ø30
20
20
15
30
15
15
30
15
120

EX-35

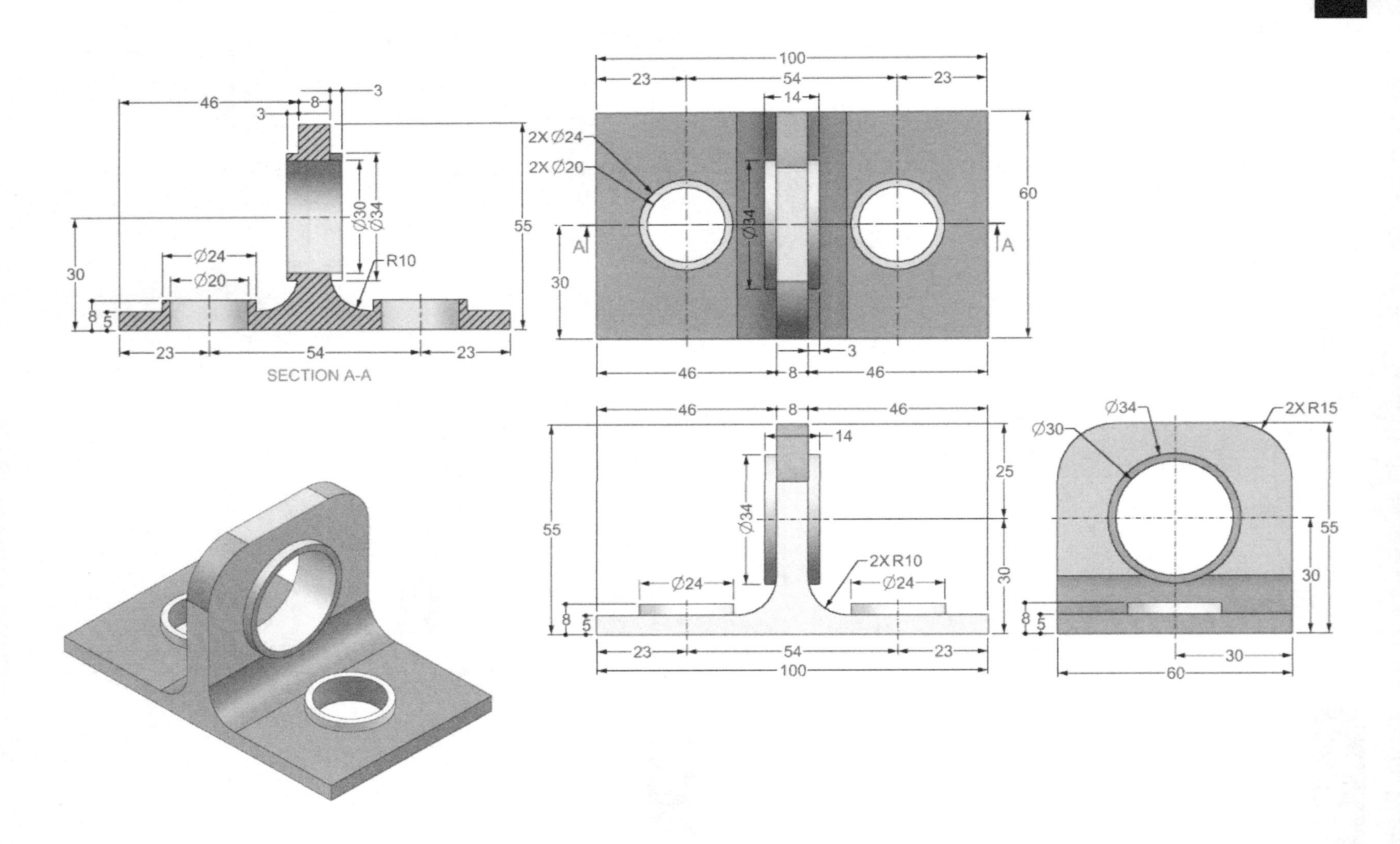

EX-36

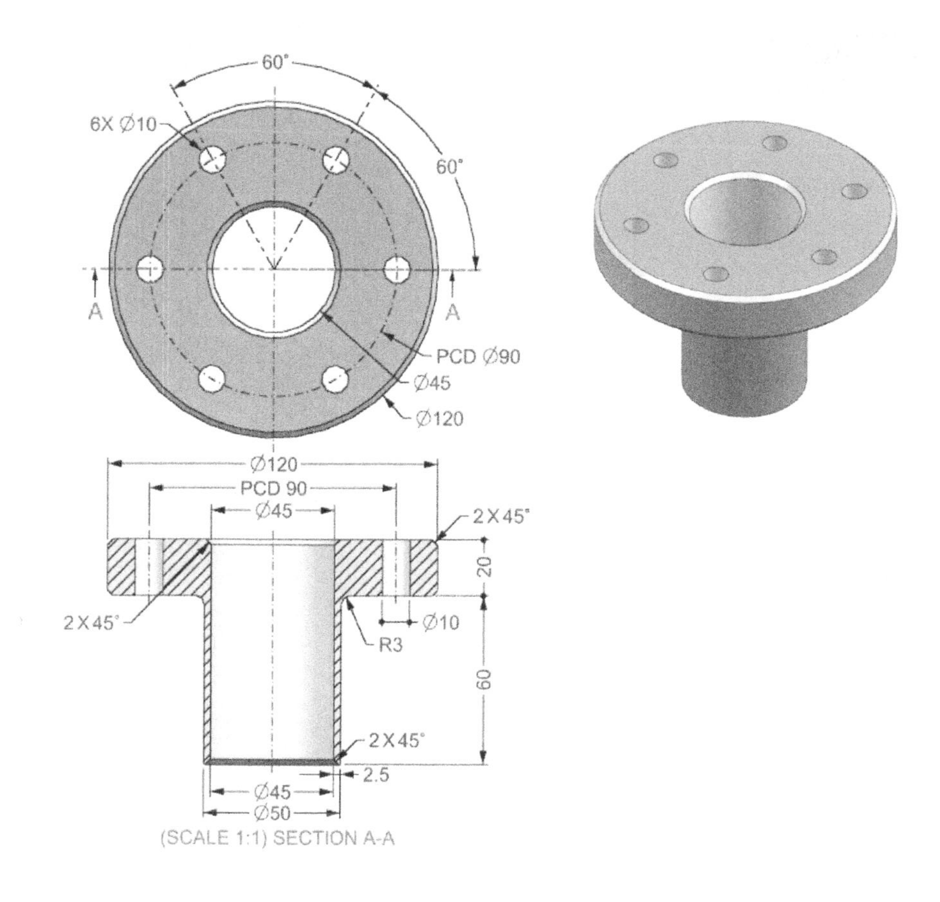
60°
6X Ø10
60°
A
A
PCD Ø90
Ø45
Ø120
Ø120
PCD 90
Ø45
2 X 45°
20
2 X 45°
Ø10
R3
60
2 X 45°
2.5
Ø45
Ø50
(SCALE 1:1) SECTION A-A

EX-37

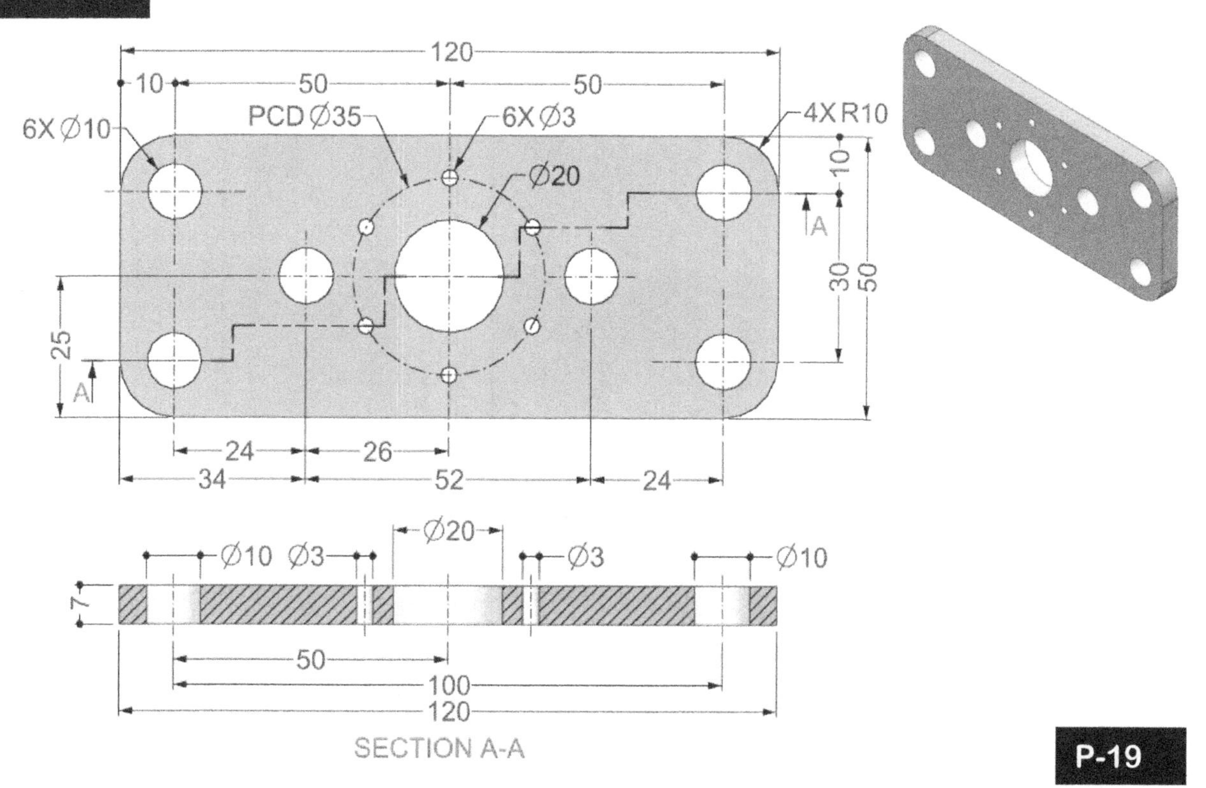
120
10
50
50
6X Ø10
PCD Ø35
6X Ø3
4X R10
Ø20
10
A
30
50
25
A
24
26
34
52
24
Ø20
Ø10
Ø3
Ø3
Ø10
7
50
100
120
SECTION A-A

EX-38

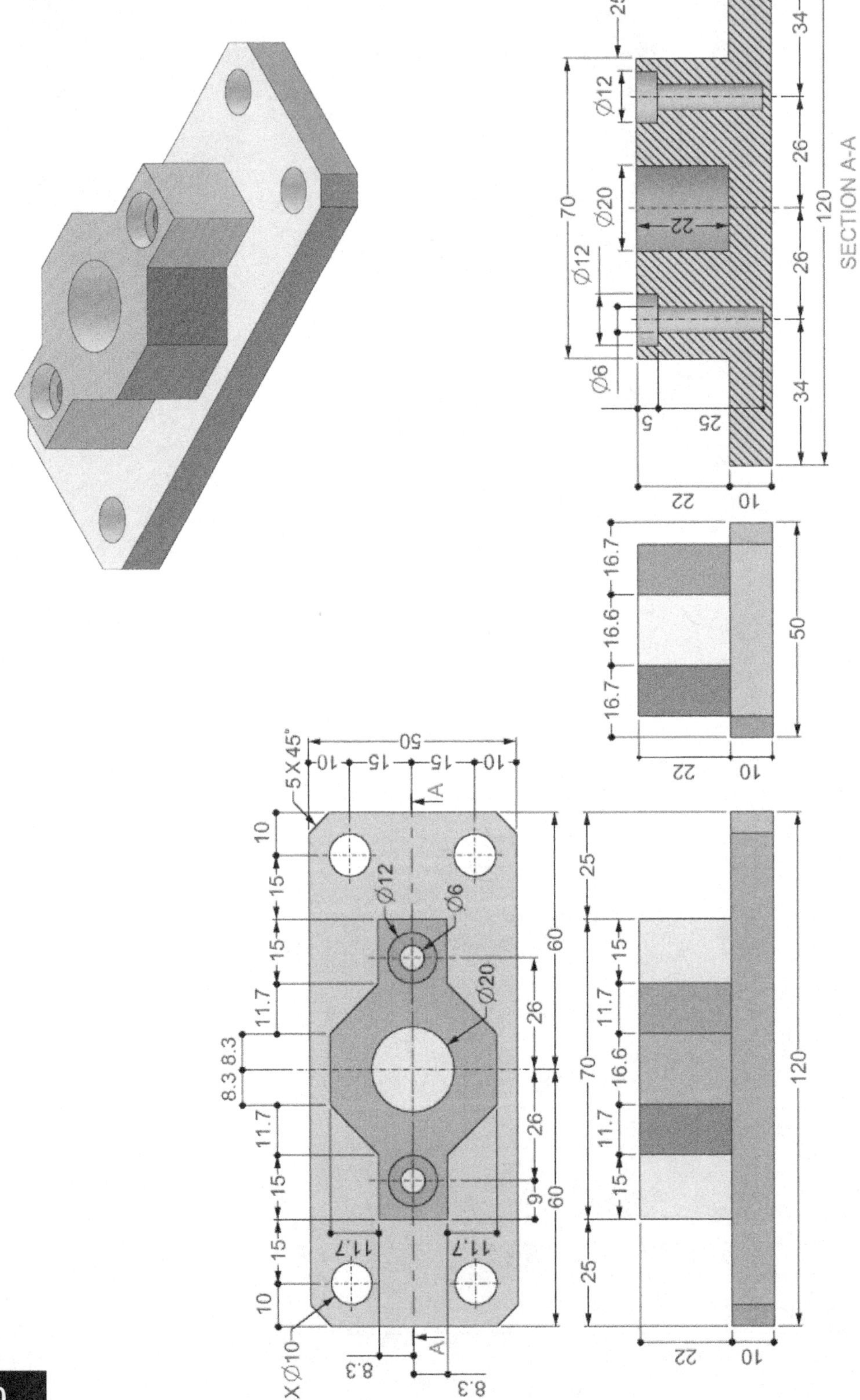

EX-39

70
R20
Ø20
40
45
45
R25
Ø20
10
20
30
A
A
10
45
65
20
2X R10
Ø40
Ø20
25
45
SECTION A-A

EX-40

Ø60
20
10
5
Ø50
Ø60
Ø50
5
10
5
30
Ø60
20

EX-41

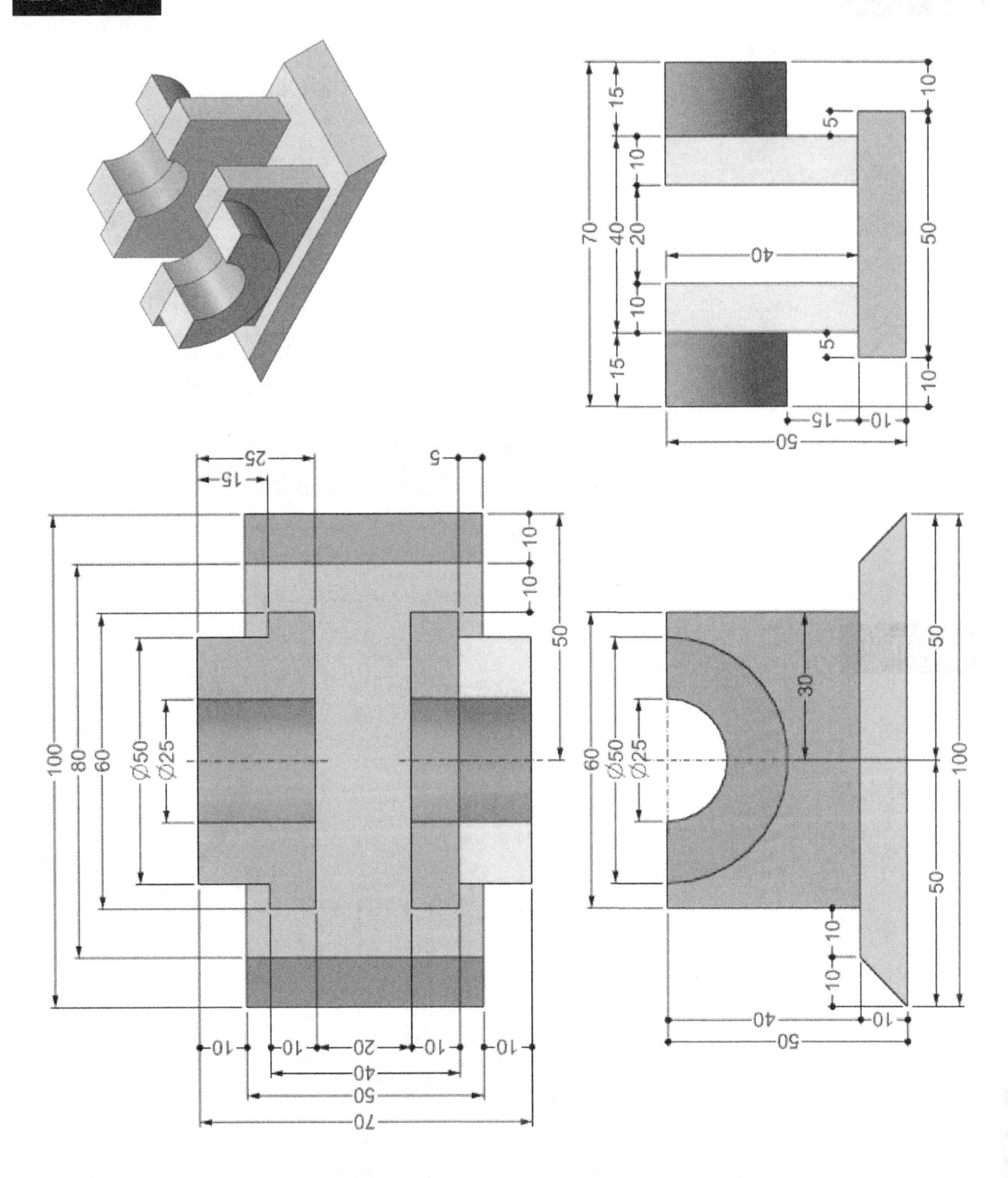

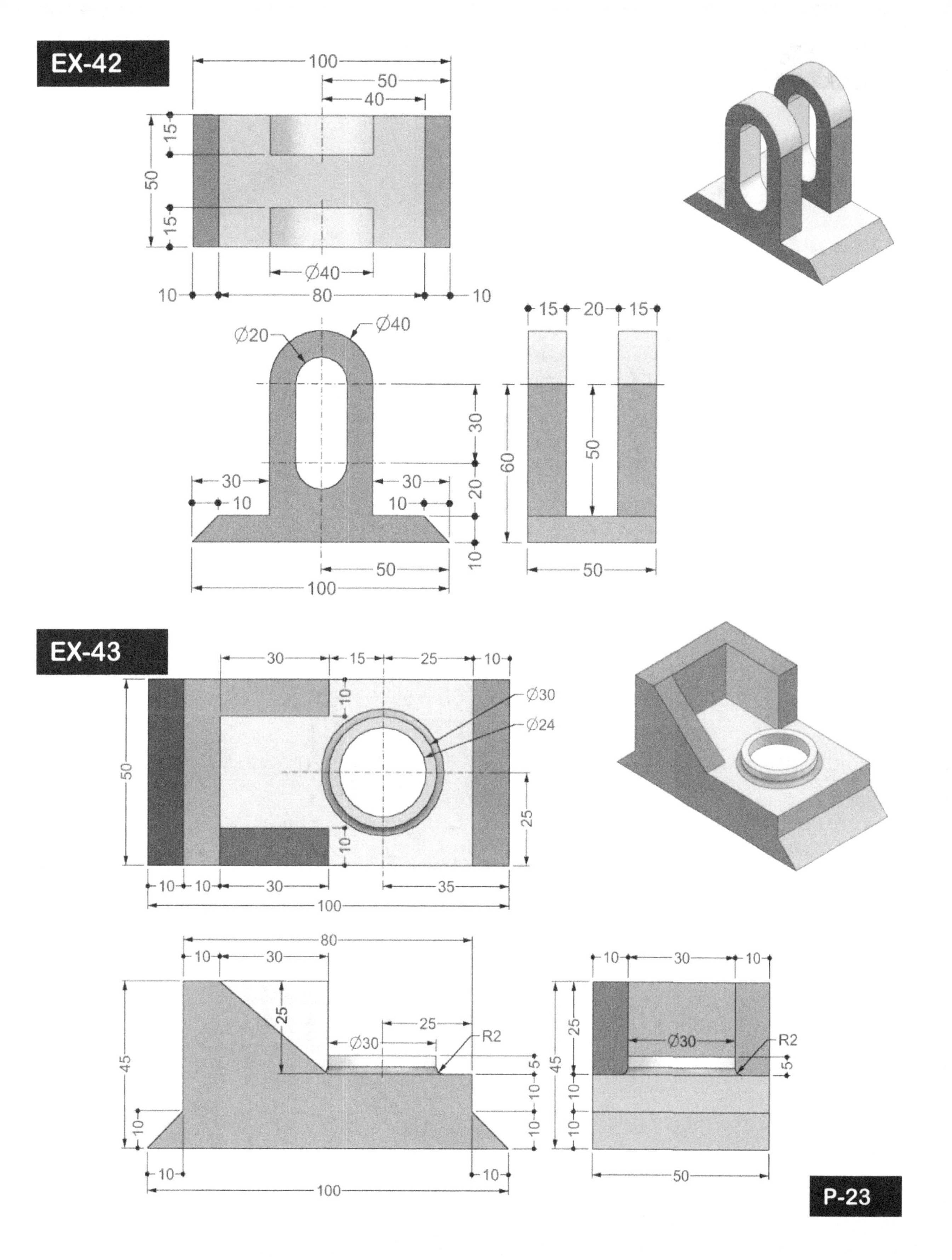

EX-42
100
50
40
15
50
15
Ø40
10
80
10
15
20
15
Ø20
Ø40
30
30
30
20
10
10
10
60
50
50
100
50
EX-43
30
15
25
10
10
Ø30
Ø24
50
25
10
10
10
10
30
35
100
80
10
30
25
25
Ø30
R2
45
10
5
10
10
10
10
100
10
30
10
25
Ø30
R2
45
5
10
10
50
P-23

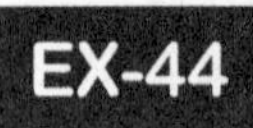

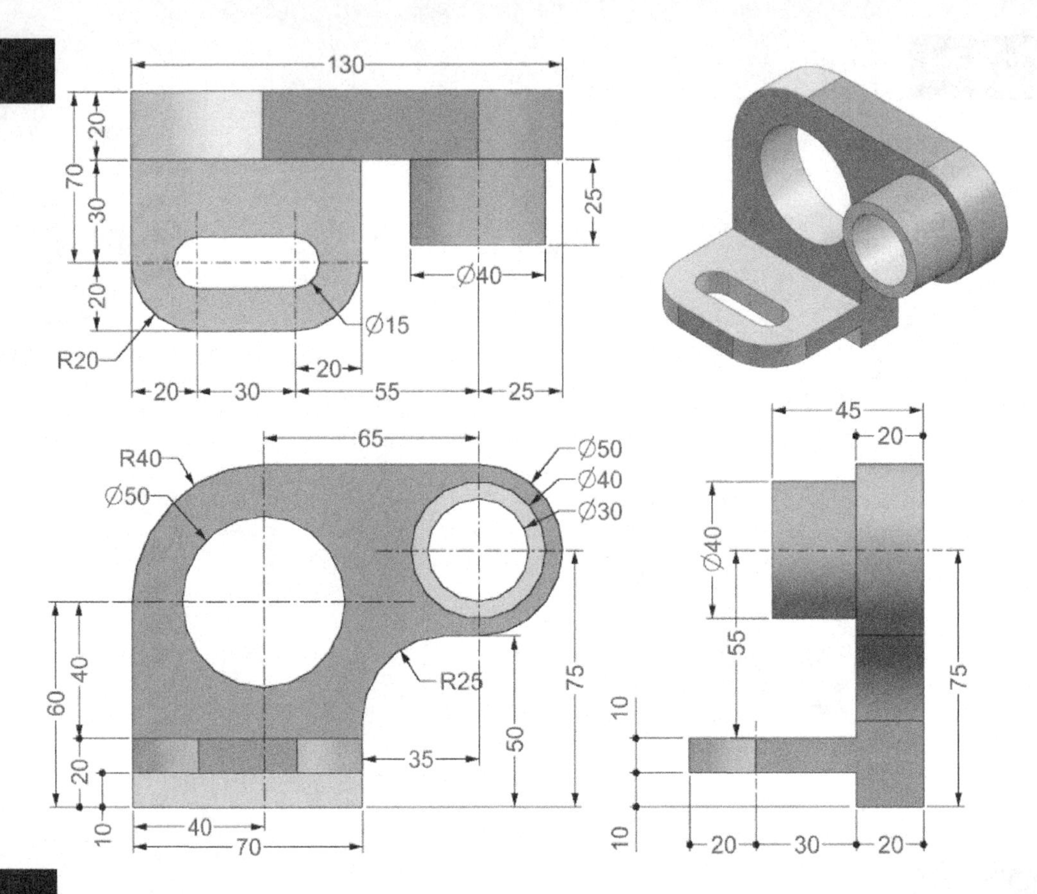
130
20
70
30
25
20
Ø40
Ø15
R20
20
20
30
55
25
45
20
65
R40
Ø50
Ø50
Ø40
Ø30
Ø40
55
75
R25
75
60
40
50
10
20
35
10
40
70
10
20
30
20

EX-45

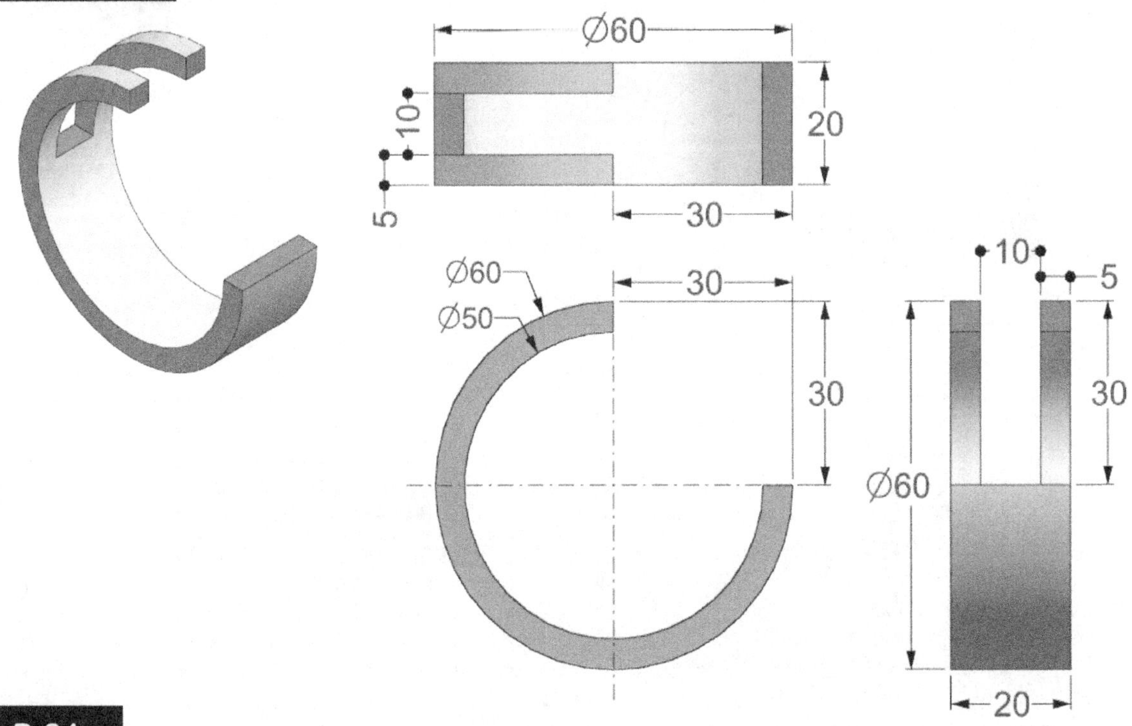
Ø60
10
20
5
30
Ø60
Ø50
30
10
5
30
30
Ø60
20

EX-46

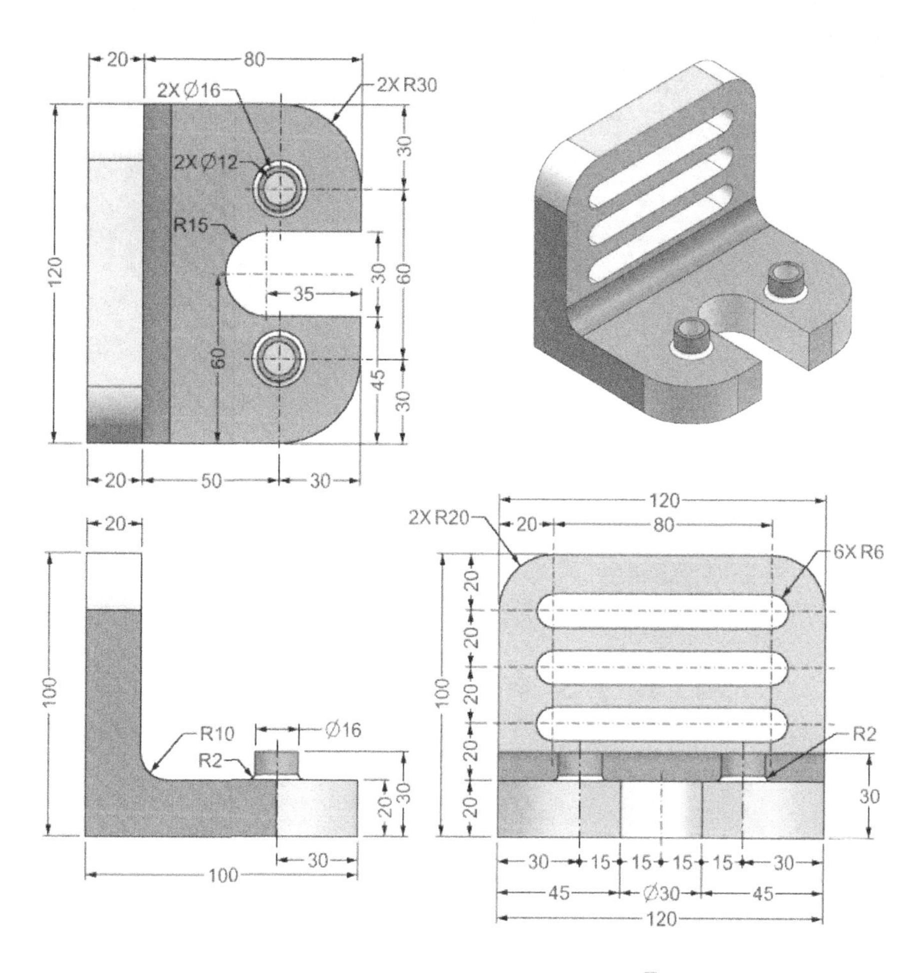

20
80
2X Ø16
2X R30
2X Ø12
R15
30
60
35
45
120
20
50
30
100
R10
R2
Ø16
2X R20
120
80
6X R6
R2
15
45
Ø30

EX-47

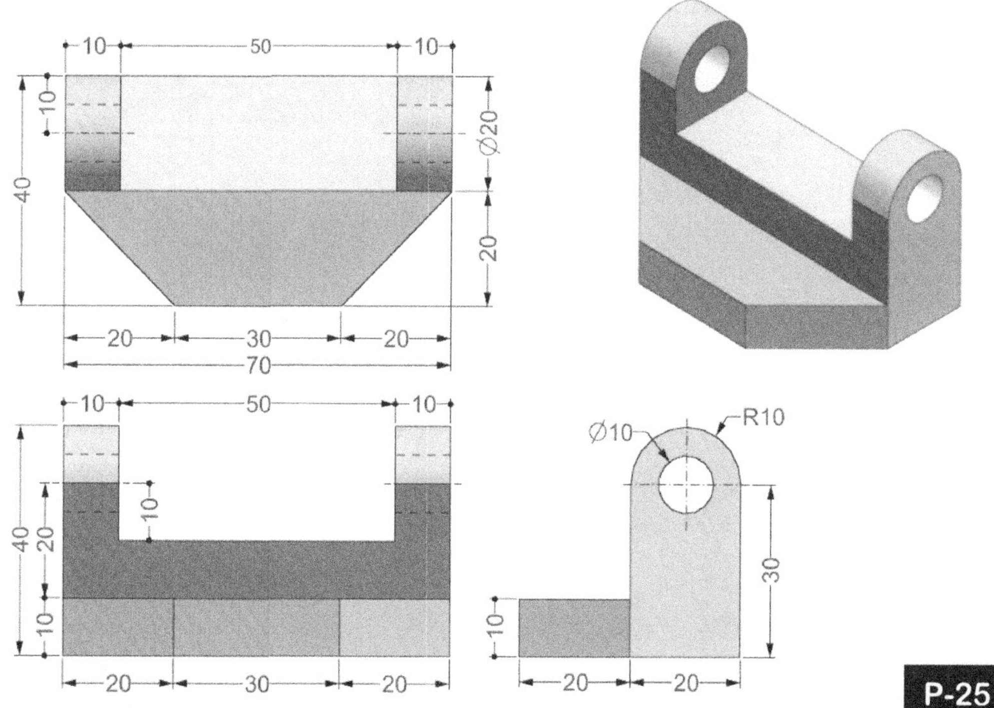

10
50
Ø20
40
20
30
70
Ø10
R10

EX-48

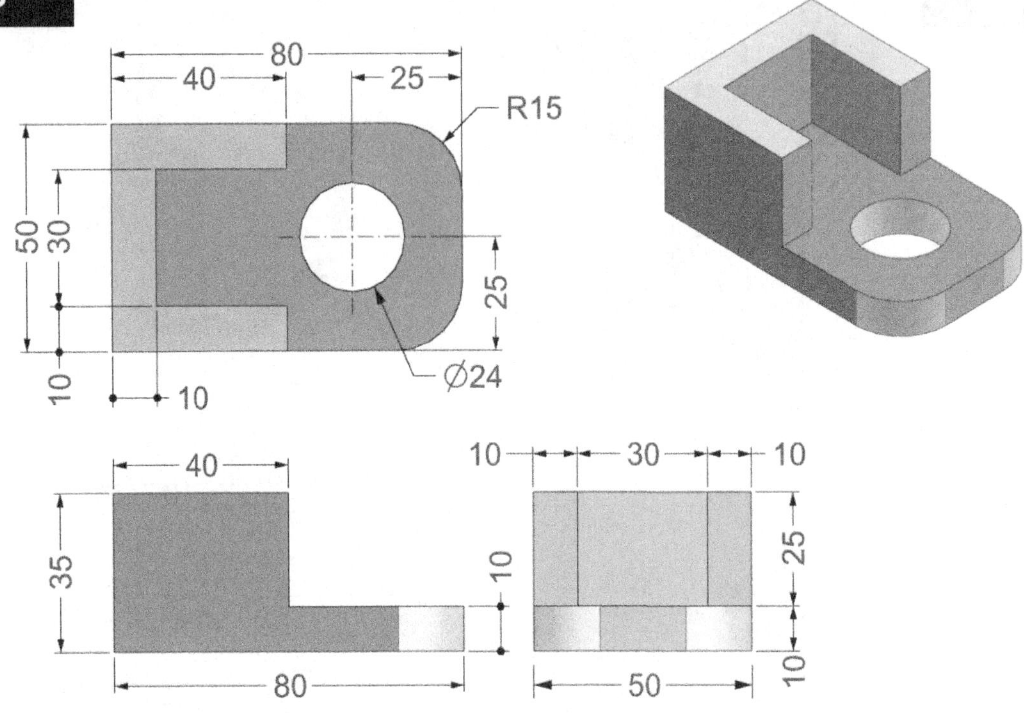
80
40
25
R15
50
30
25
Ø24
10
10
40
35
80
10
10
30
10
25
10
50

EX-49

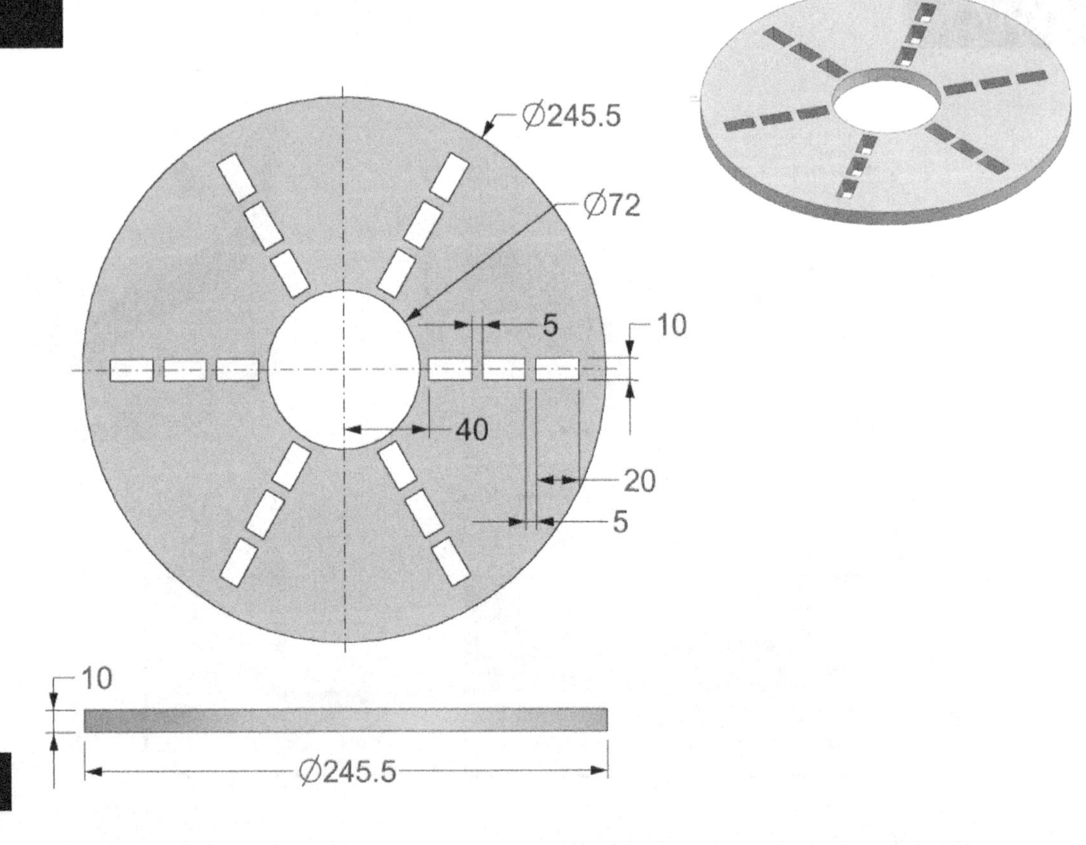
Ø245.5
Ø72
5
10
40
20
5
10
Ø245.5

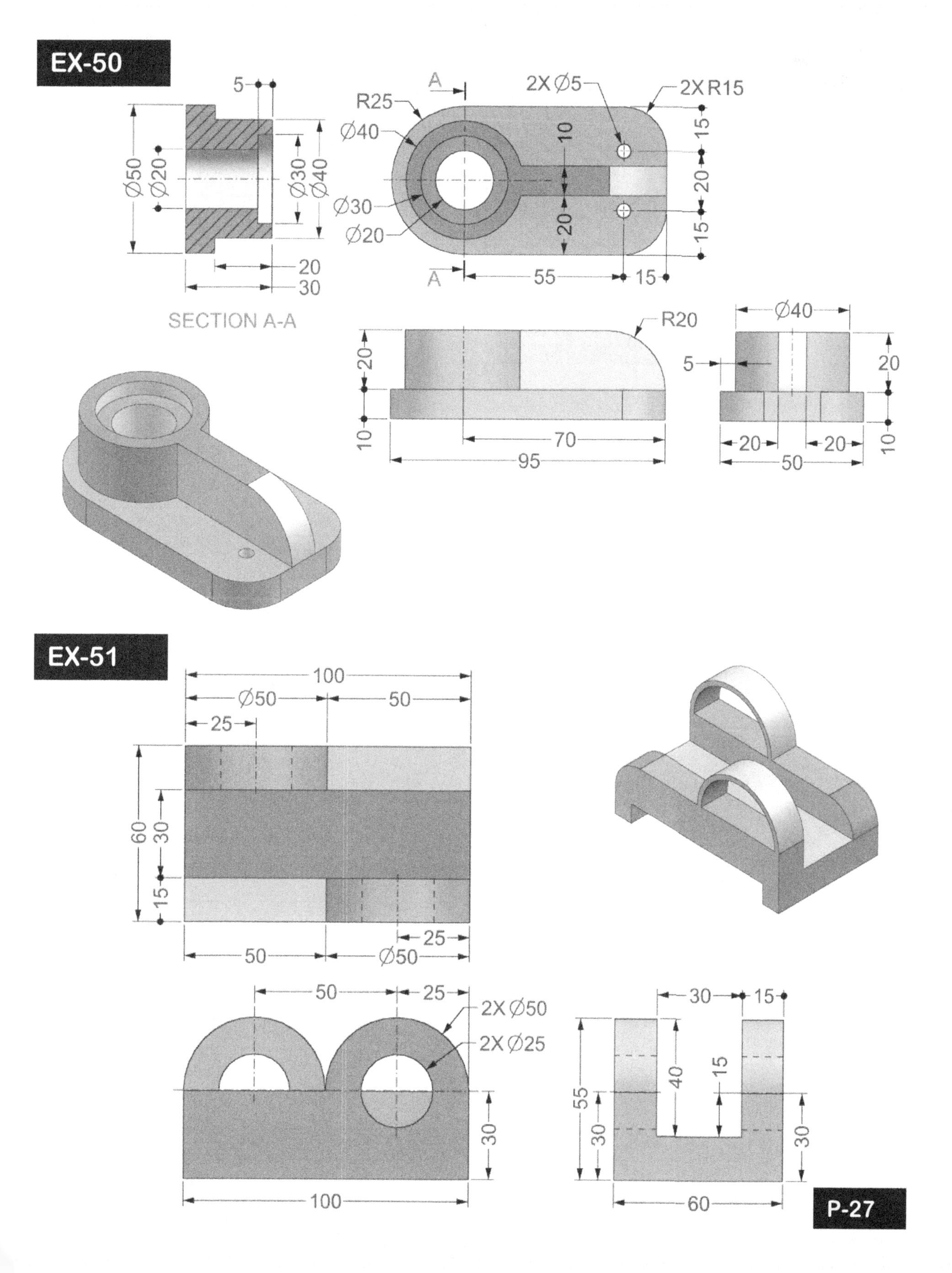
EX-50
SECTION A-A
A
A
5
Ø50
Ø20
Ø30
Ø40
20
30
R25
Ø40
Ø30
Ø20
2X Ø5
2X R15
10
20
55
15
R20
70
95
Ø40
5
20
10
50
EX-51
100
Ø50
50
25
60
30
15
2X Ø50
2X Ø25
55
40
60
P-27

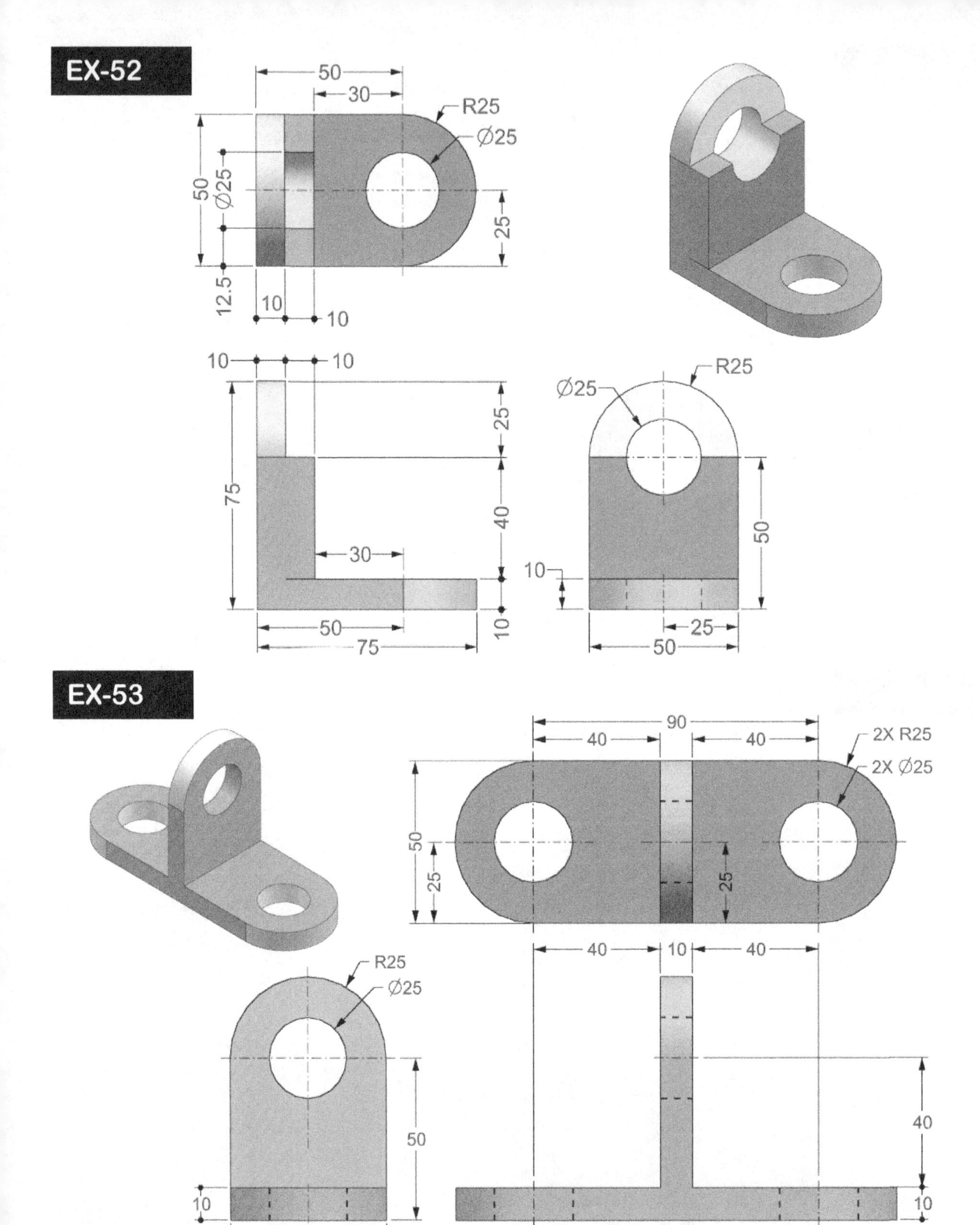

EX-52
50
30
R25
∅25
50
∅25
25
12.5
10
10
10
10
25
75
40
30
10
50
75
R25
∅25
10
50
25
50
EX-53
90
40
40
2X R25
2X ∅25
50
25
25
40
10
40
R25
∅25
50
10
25
50
40
10
90

EX-54

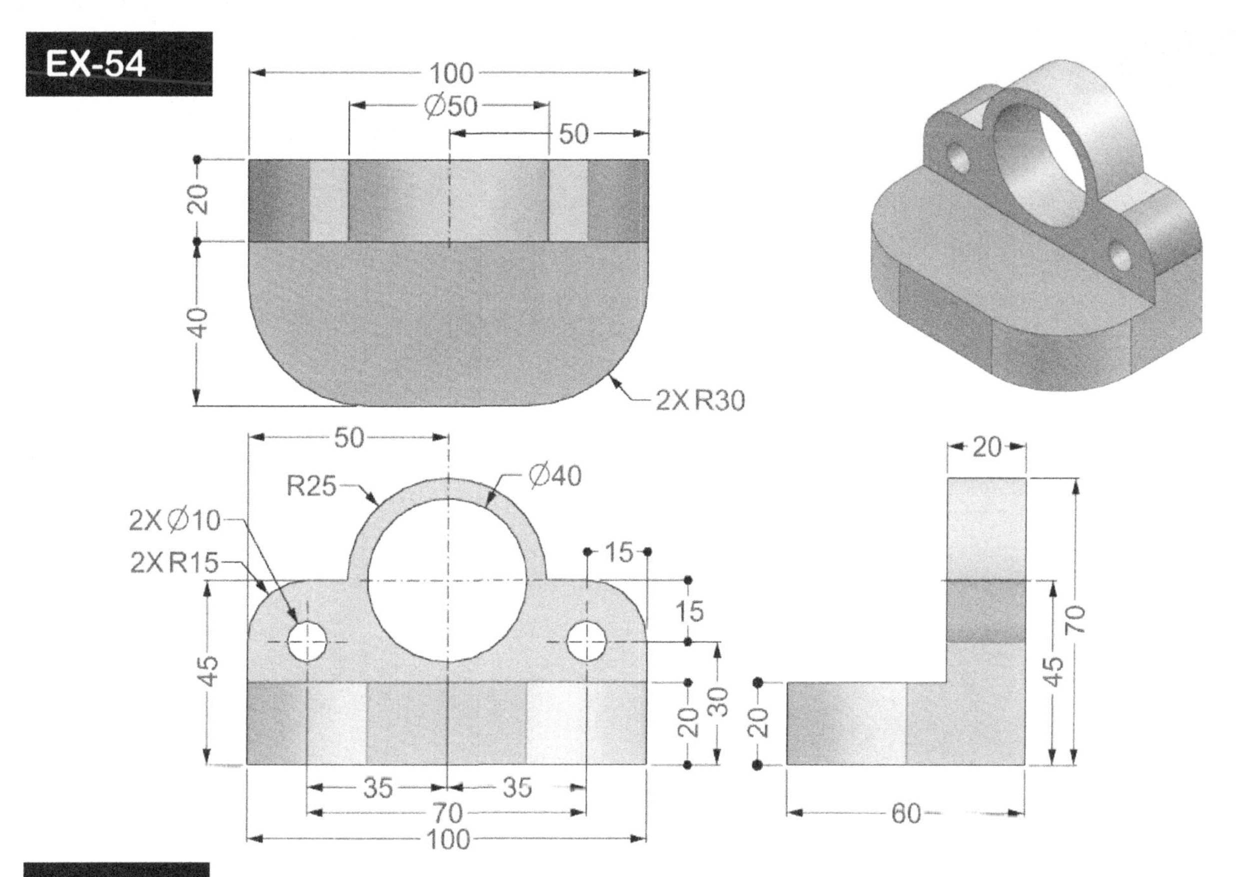

100
∅50
50
20
40
2X R30
50
R25
∅40
2X∅10
2X R15
15
15
45
20
30
20
35
35
70
100
20
70
45
60

EX-55

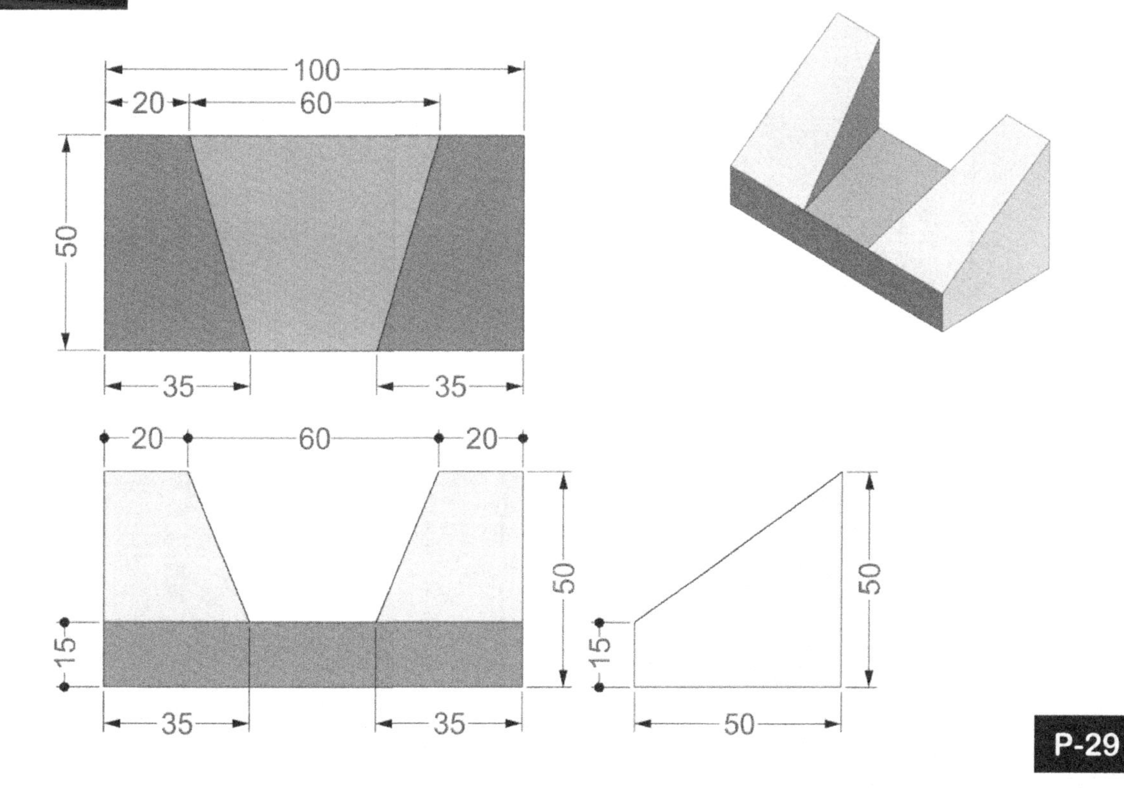

100
20
60
50
35
35
20
60
20
50
15
15
50
35
35
50

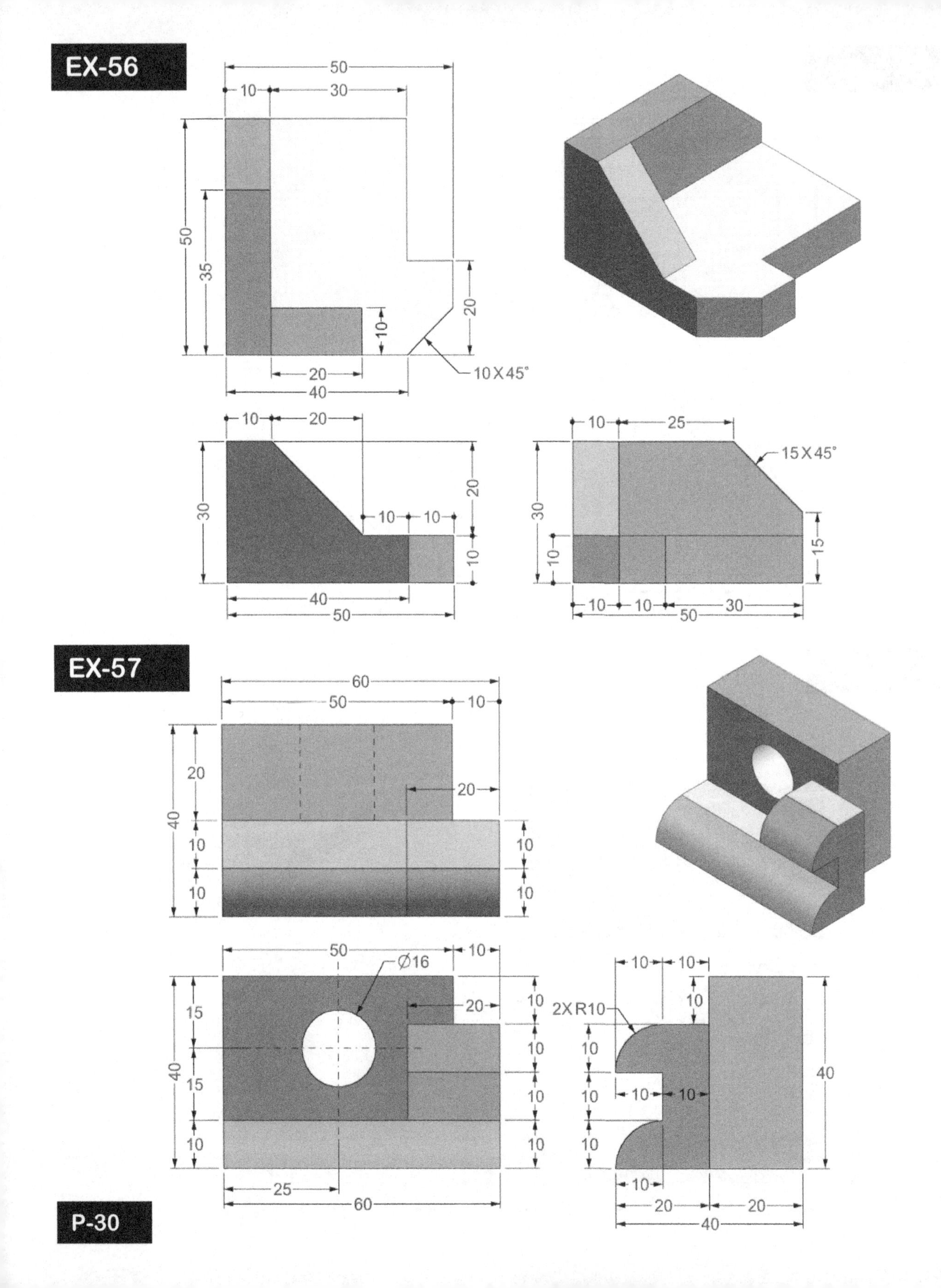

EX-56
10X45°
15X45°
EX-57
Ø16
2XR10
P-30

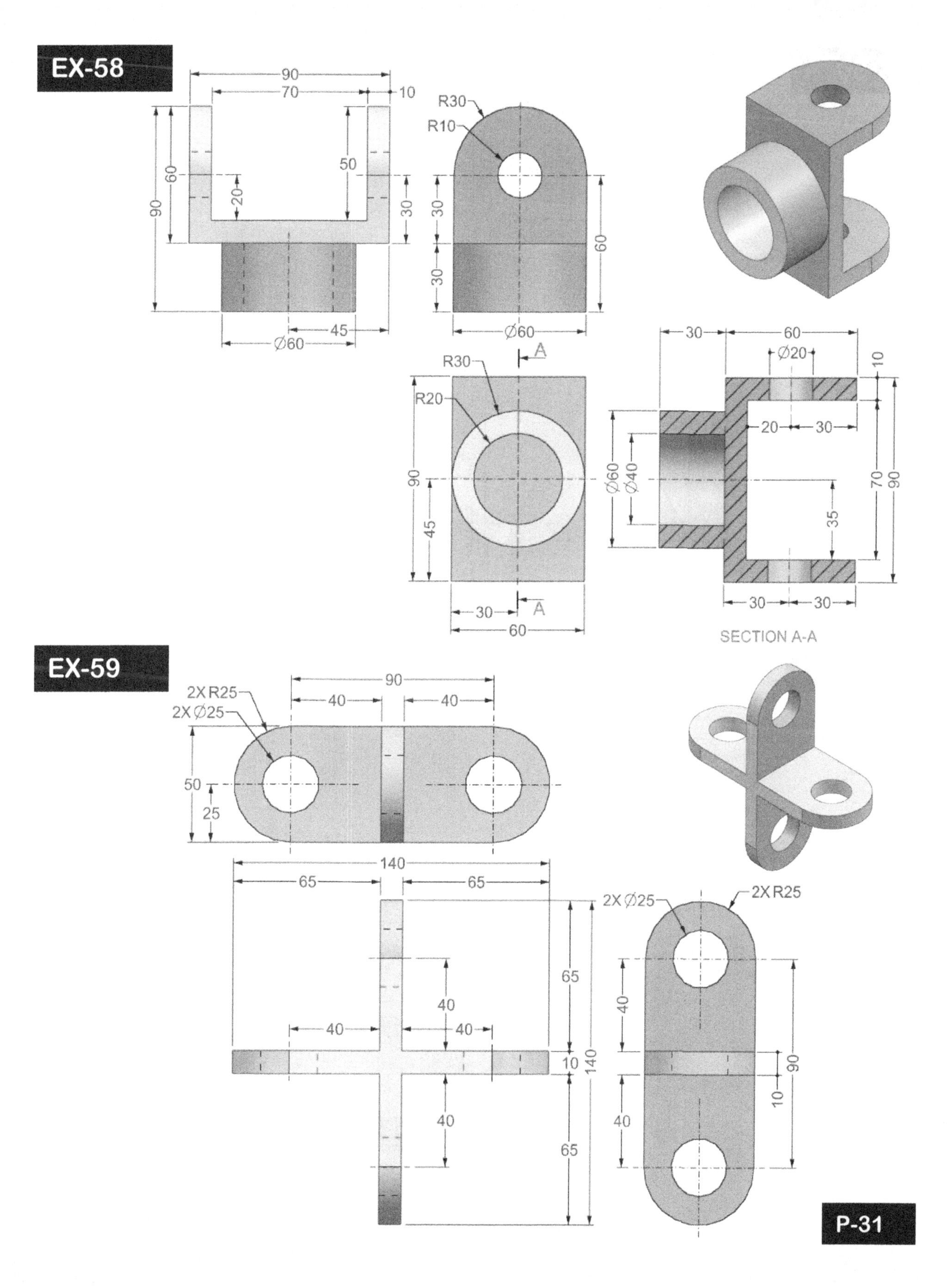

EX-58
90
70
10
R30
R10
60
50
20
30
30
30
60
90
45
Ø60
Ø60
R30
R20
A
A
90
45
30
60
30
60
Ø20
10
20
30
Ø60
Ø40
70
90
35
30
30
SECTION A-A
EX-59
90
40
40
2X R25
2X Ø25
50
25
140
65
65
40
40
40
65
10
140
40
65
2X Ø25
2X R25
40
90
10
40
P-31

EX-60

EX-61

SECTION A-A

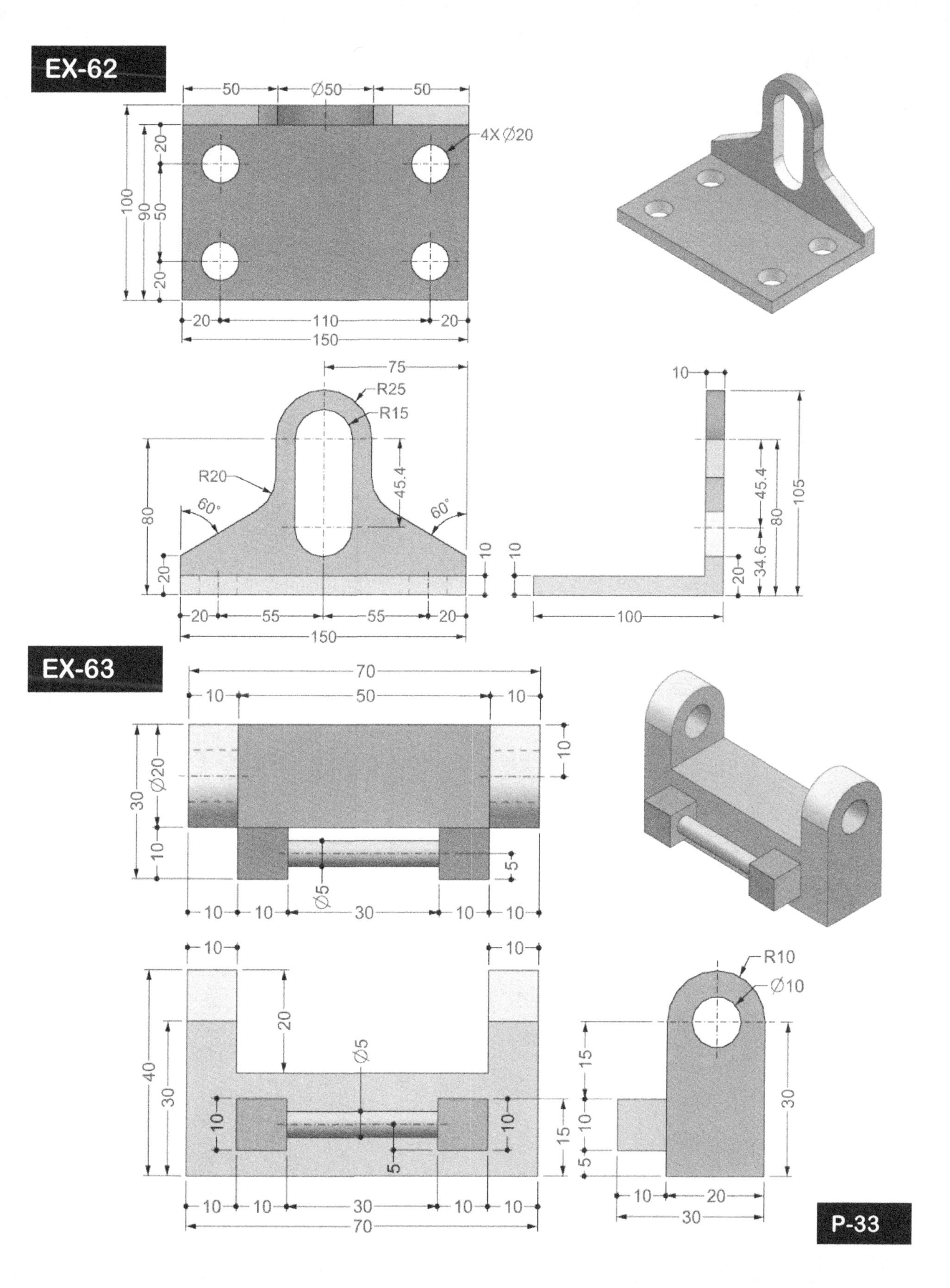
EX-62
50
Ø50
50
4X Ø20
100
90
20
50
20
20
110
20
150
75
R25
R15
R20
60°
45.4
60°
80
20
10
20
55
55
20
150
10
10
45.4
105
80
34.6
20
100
EX-63
70
10
50
10
10
30
Ø20
10
5
Ø5
10
10
30
10
10
10
10
R10
Ø10
40
30
20
Ø5
10
10
15
15
10
5
5
30
10
10
30
10
10
70
10
20
30
P-33

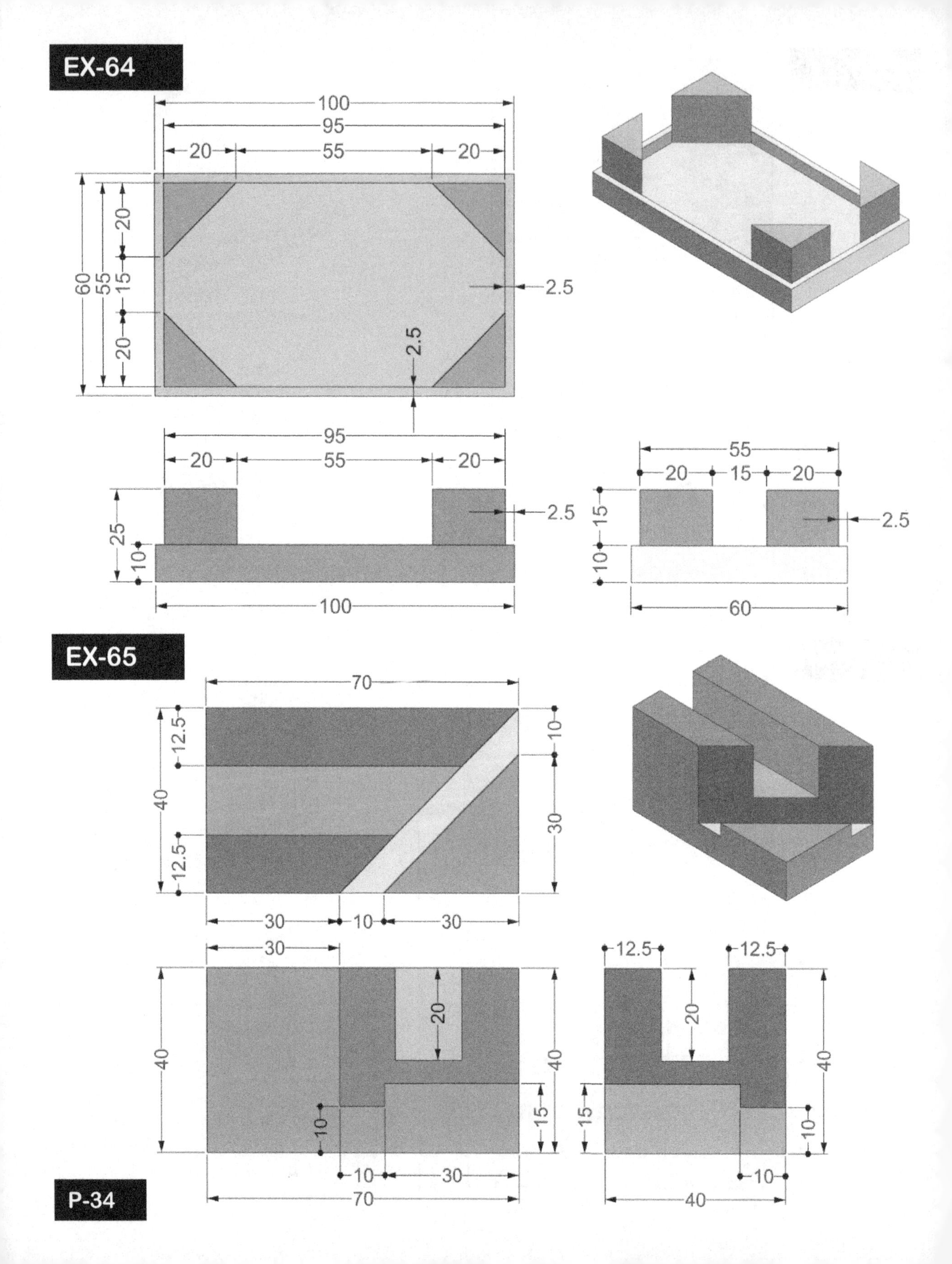

EX-64
EX-65
P-34

EX-66

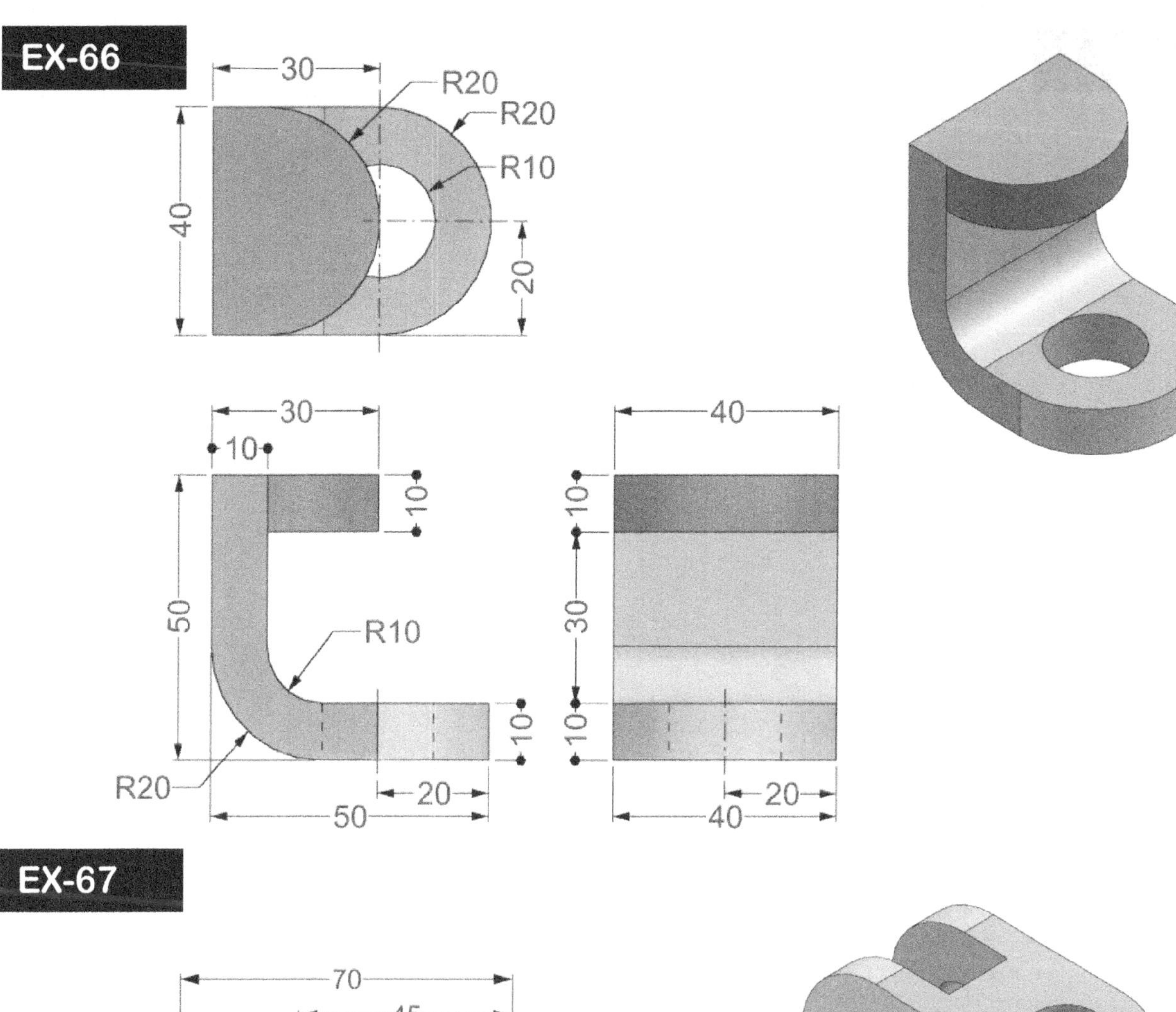

30
R20
R20
R10
40
20
30
10
10
50
R10
10
R20
20
50
40
10
30
10
20
40

EX-67

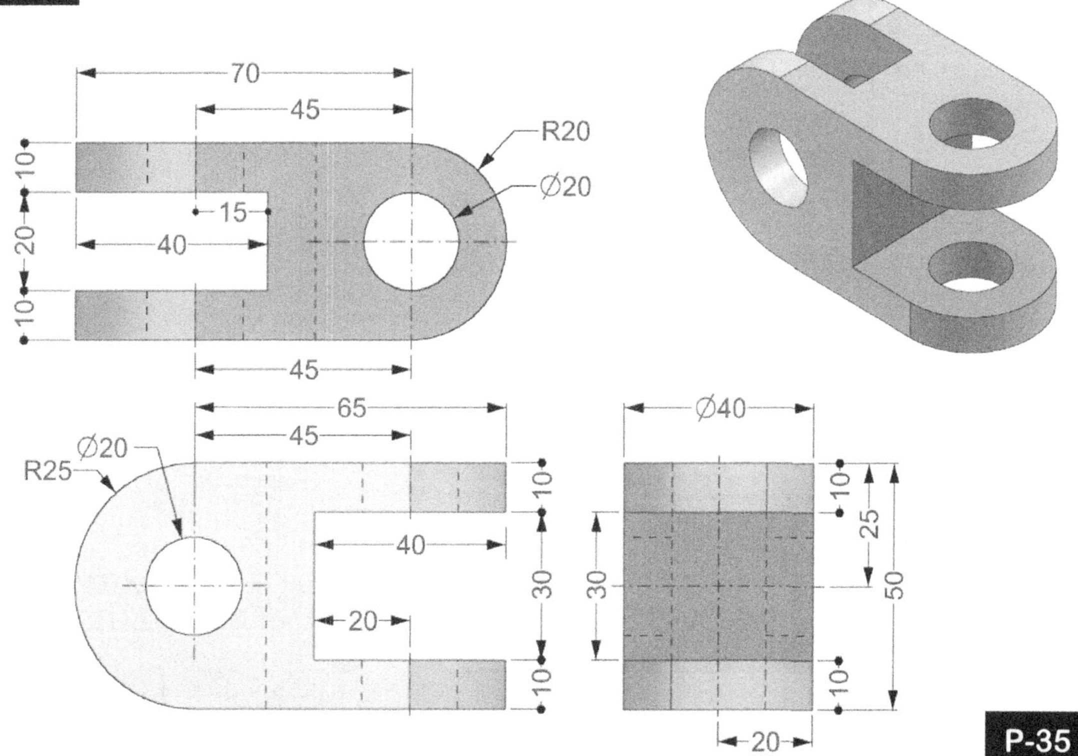

70
45
R20
Ø20
10
15
20
40
10
45
65
45
Ø20
R25
10
40
30
20
10
Ø40
10
25
30
50
10
20

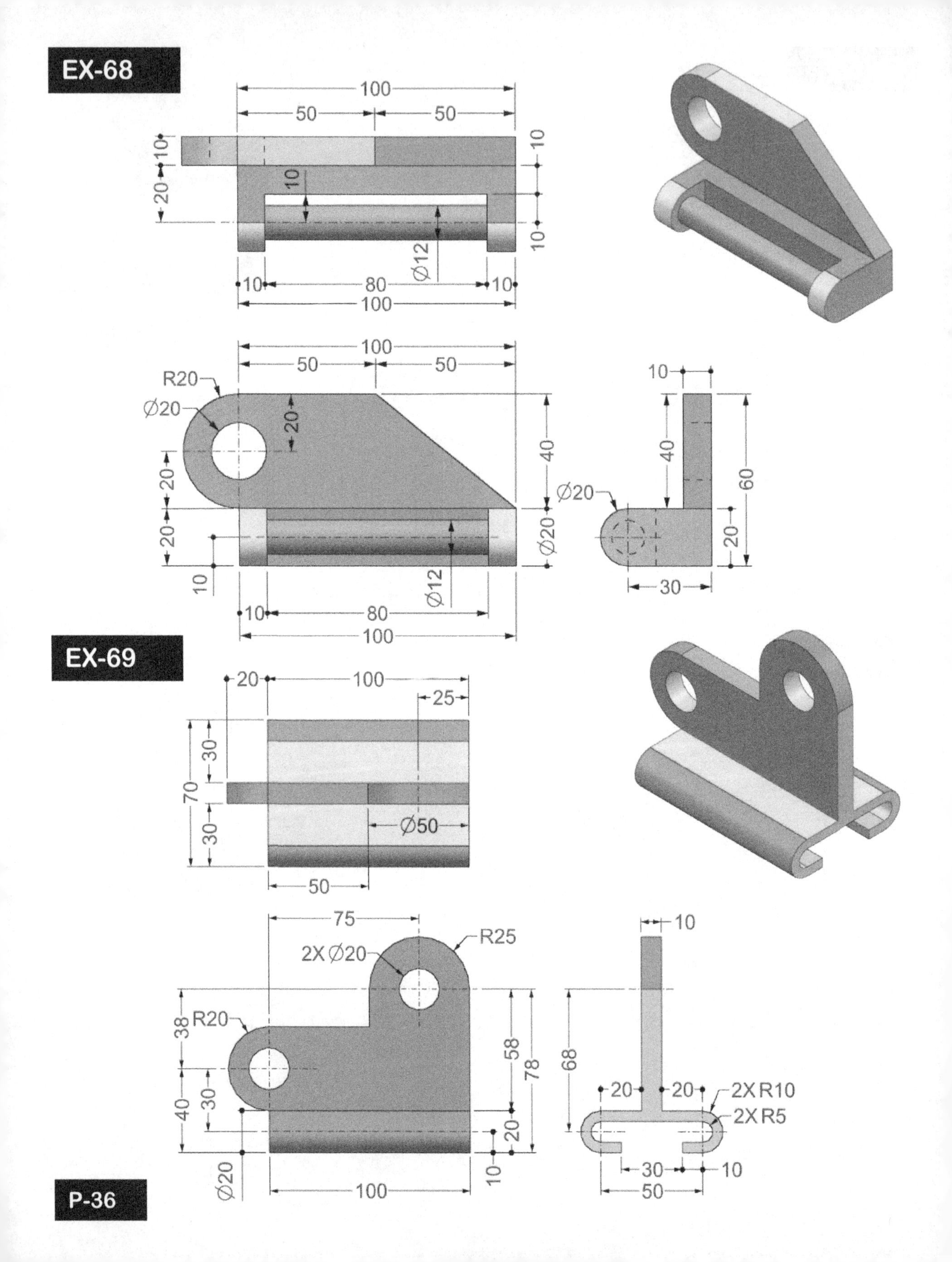

EX-68
100
50
50
10
20
10
10
10
Ø12
10
80
10
100
100
50
50
R20
Ø20
20
40
20
20
10
Ø20
Ø12
10
80
100
10
40
60
Ø20
20
30
EX-69
20
100
25
30
70
30
Ø50
50
75
2X Ø20
R25
R20
38
30
40
58
78
20
10
Ø20
100
10
68
20
20
2X R10
2X R5
30
10
50
P-36

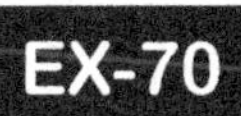

EX-70

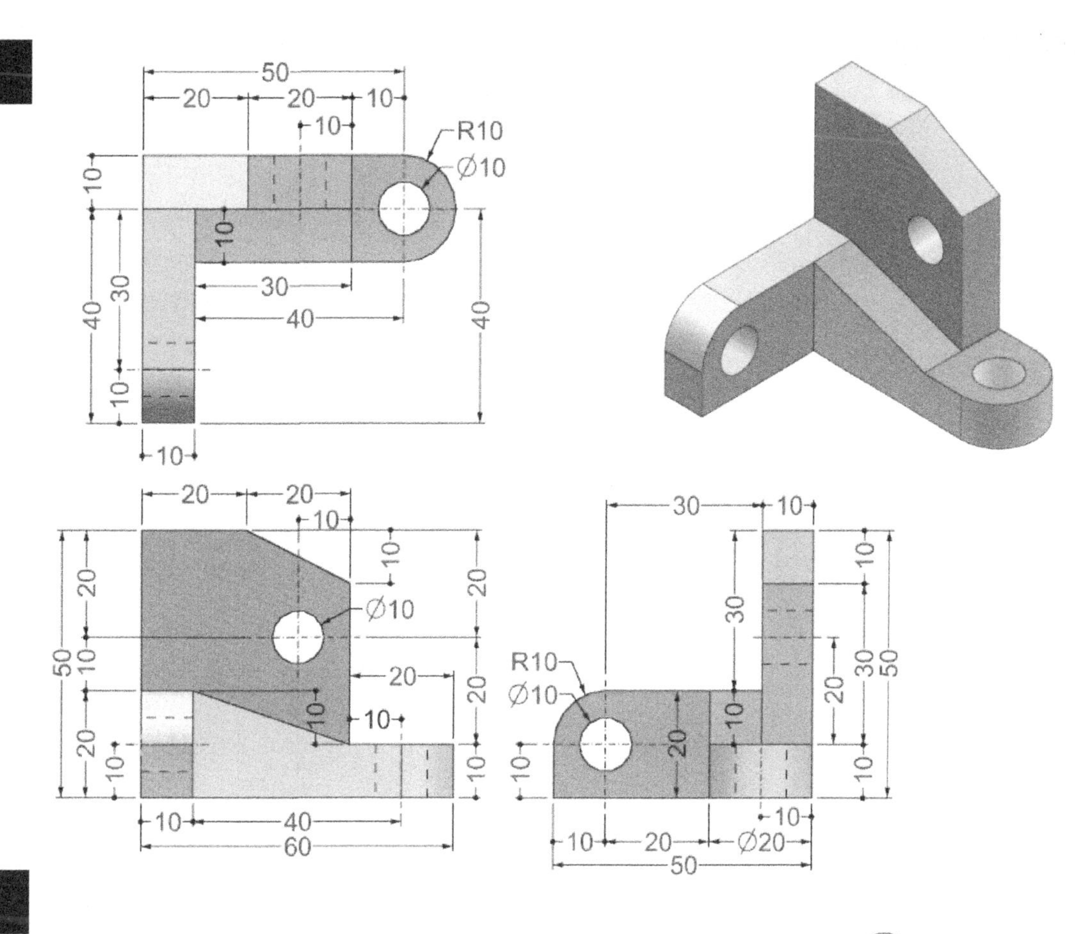

50
20
20
10
10
R10
∅10
10
10
30
40
30
40
40
10
10
20
20
10
10
20
∅10
20
50
10
20
20
10
20
10
10
10
10
40
60
30
10
R10
∅10
20
10
30
10
30
50
20
10
10
10
10
20
∅20
50

EX-71

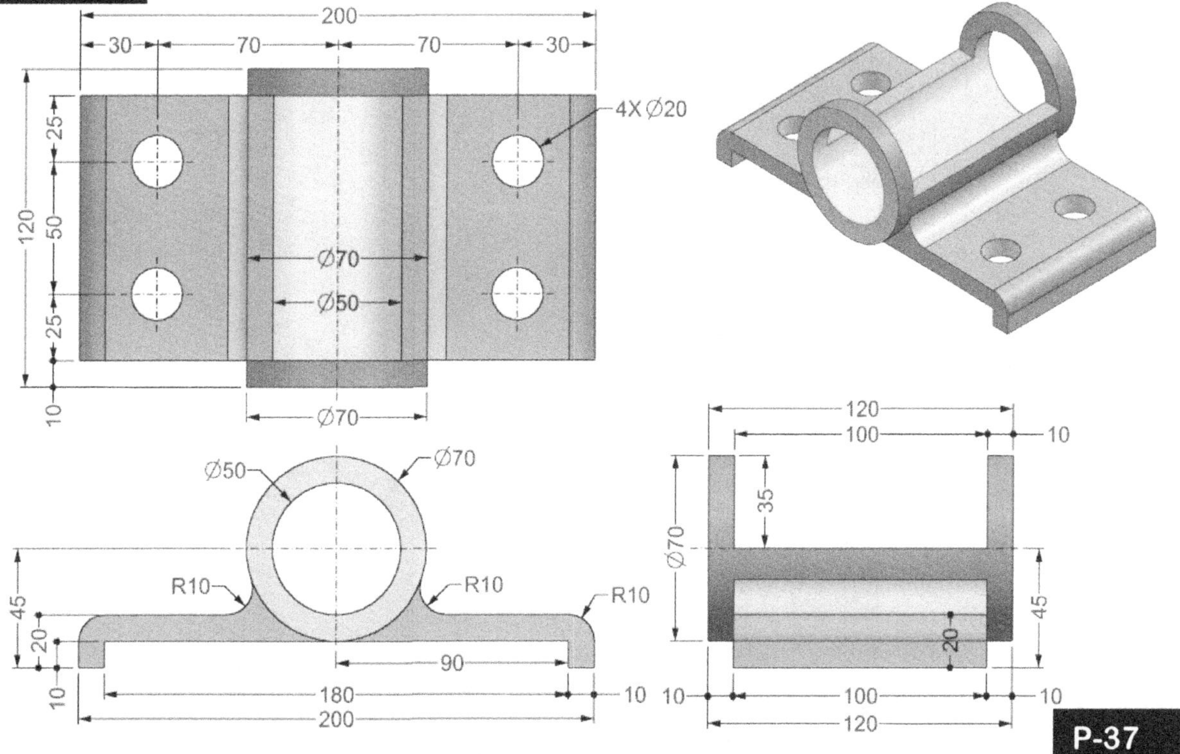

200
30
70
70
30
4X ∅20
120
25
50
25
10
∅70
∅50
∅70
∅50
∅70
R10
R10
R10
45
20
10
90
180
10
200
120
100
10
35
∅70
45
20
10
100
10
120

EX-72

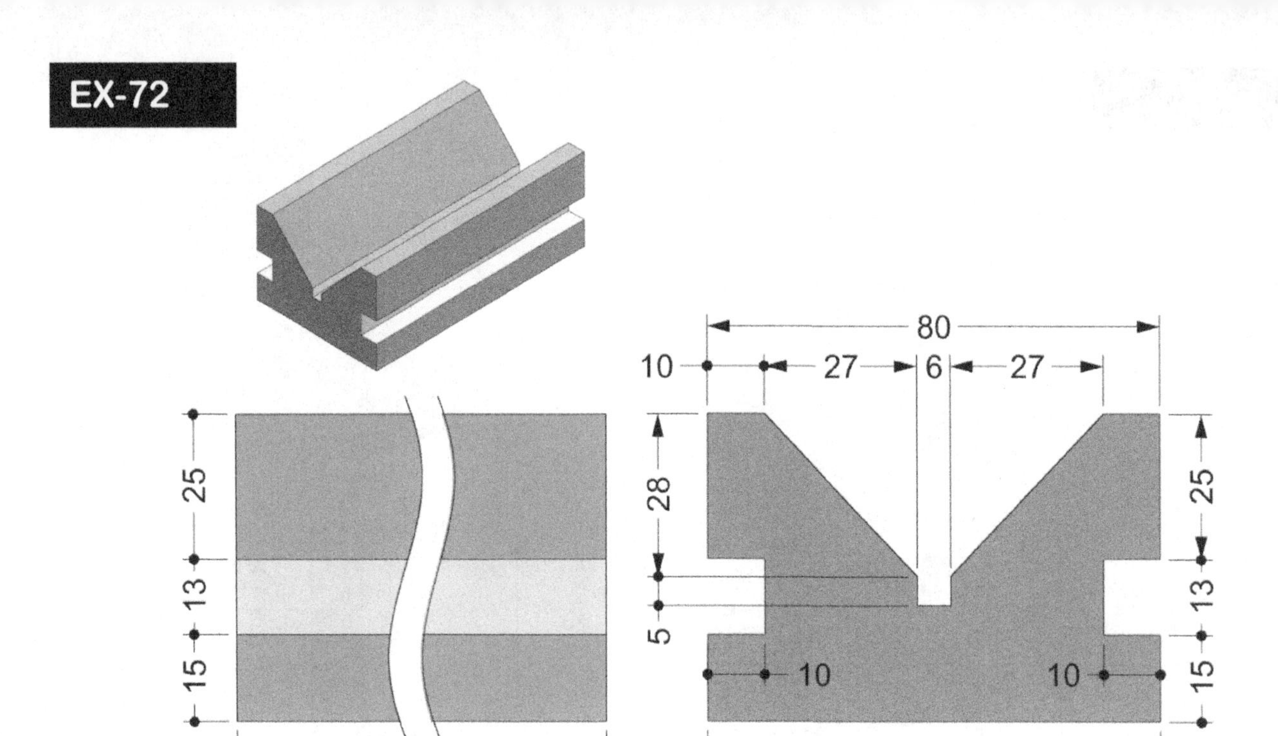

80
10
27
6
27
25
13
15
28
5
10
10
25
13
15
140
80

EX-73

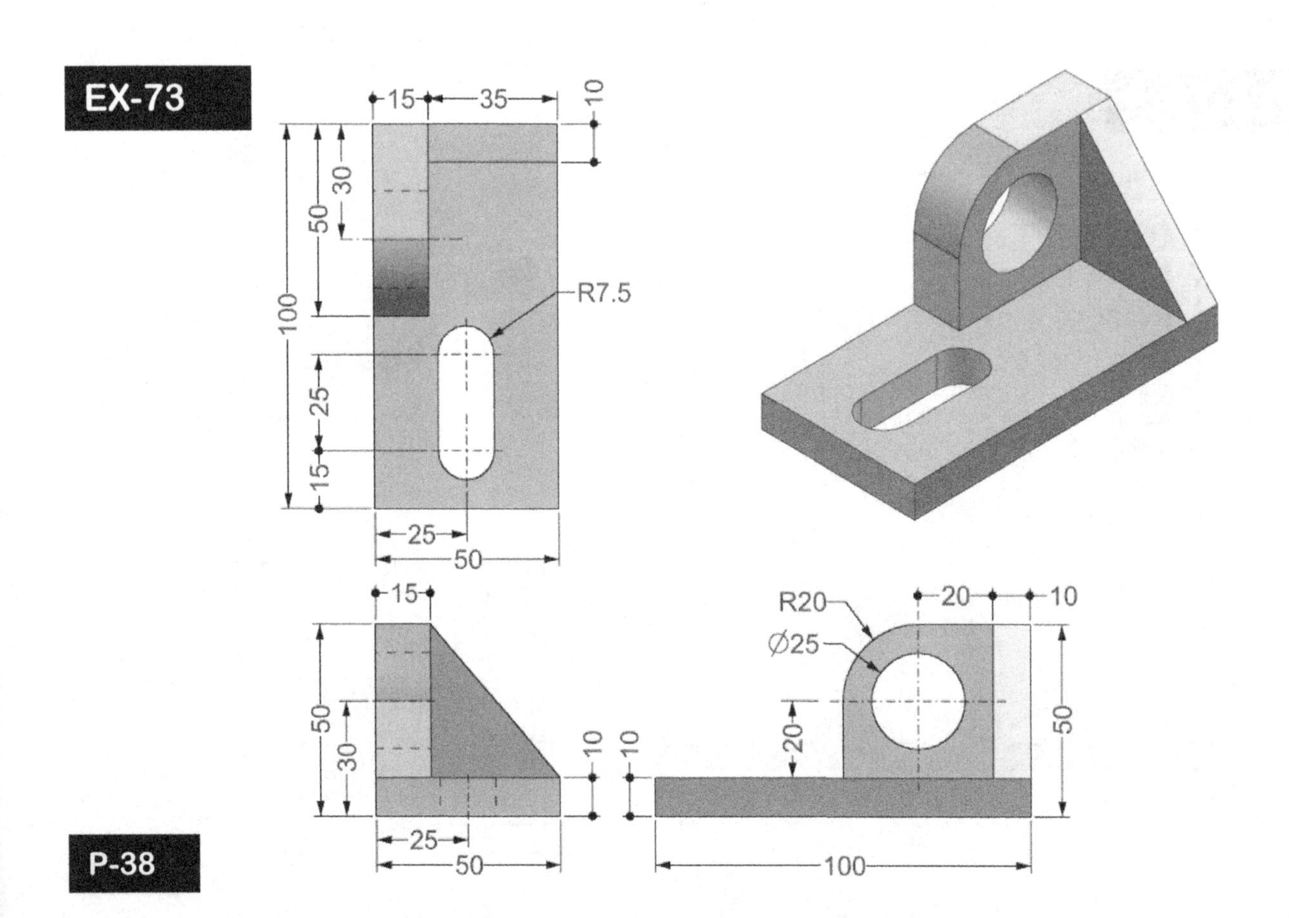

15
35
10
30
50
100
R7.5
25
15
25
50
15
R20
Ø25
20
10
50
30
20
10
10
25
50
100

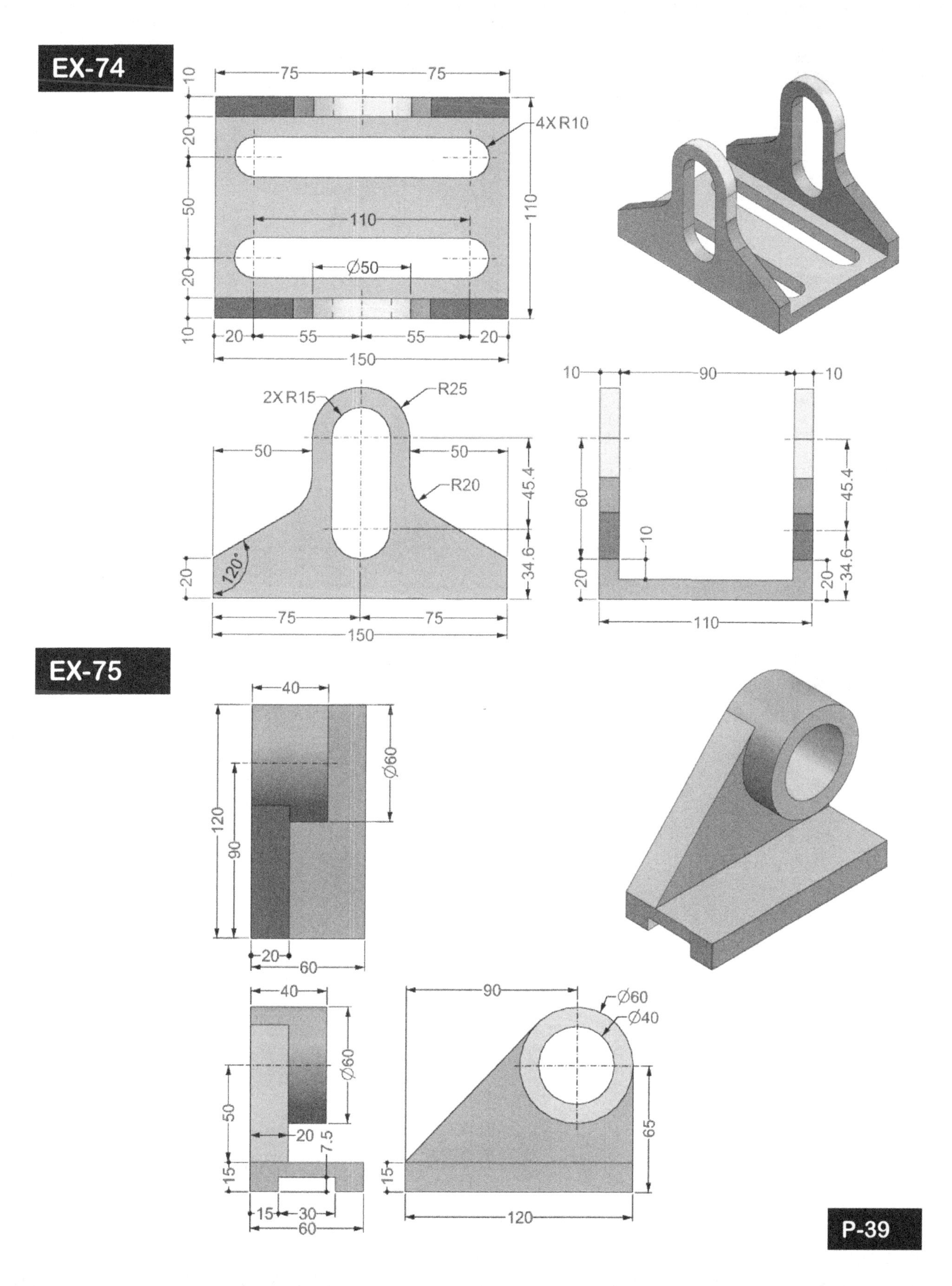

EX-74
75
75
4XR10
110
50
20
20
10
10
110
Ø50
20
55
55
20
150
2XR15
R25
50
50
R20
45.4
34.6
20
120°
75
75
150
10
90
10
60
10
20
45.4
20
34.6
110
EX-75
40
120
90
Ø60
20
60
40
Ø60
50
20
7.5
15
15
30
60
90
Ø60
Ø40
65
15
120

EX-76

EX-77

SECTION A-A

EX-78

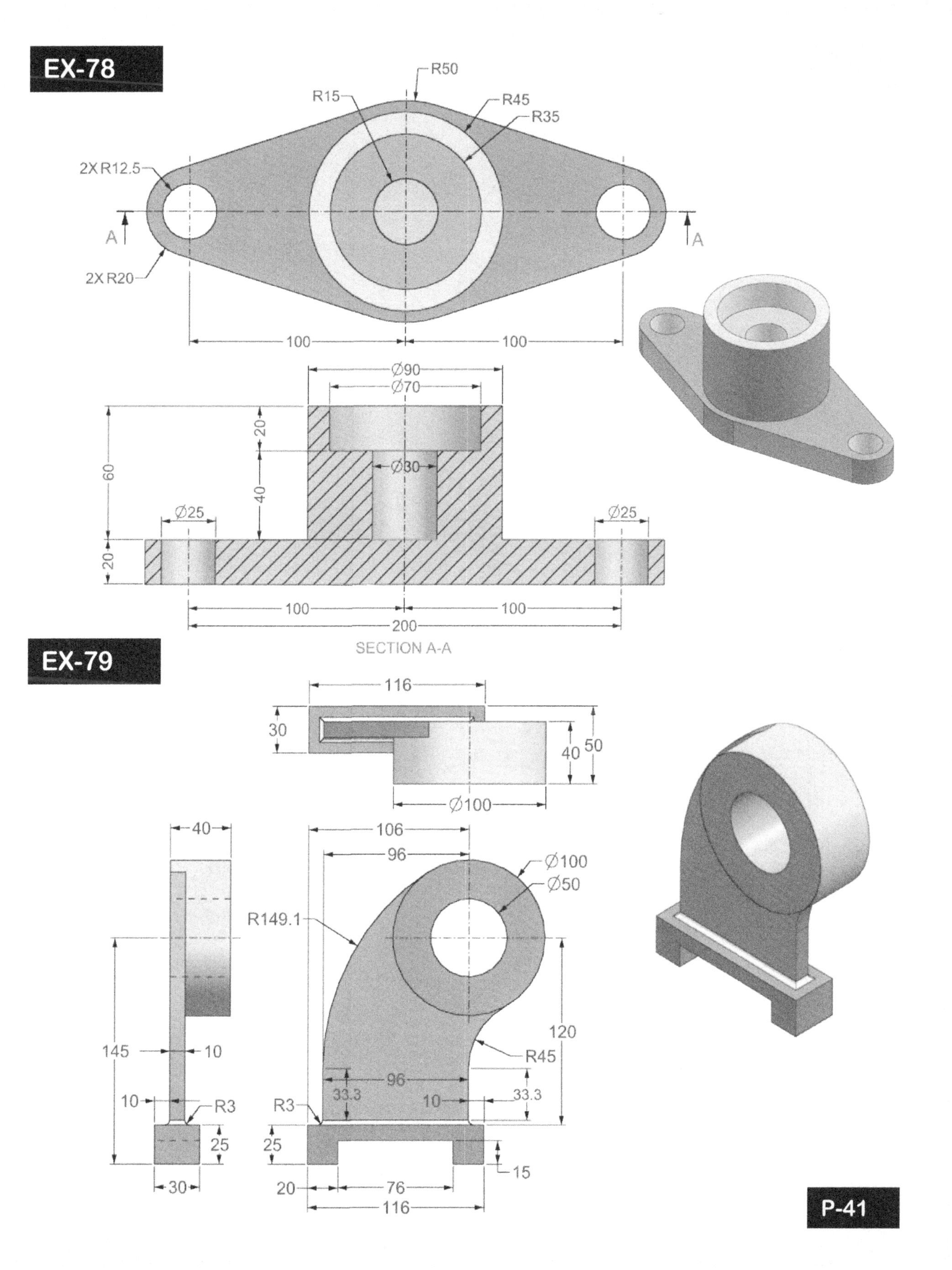

R50
R15
R45
R35
2X R12.5
A
A
2X R20
100
100
∅90
∅70
20
60
40
∅30
∅25
∅25
20
100
100
200
SECTION A-A

EX-79

116
30
40
50
∅100
40
106
96
∅100
∅50
R149.1
120
145
10
R45
96
33.3
10
33.3
10
R3
R3
25
25
15
30
20
76
116

EX-80

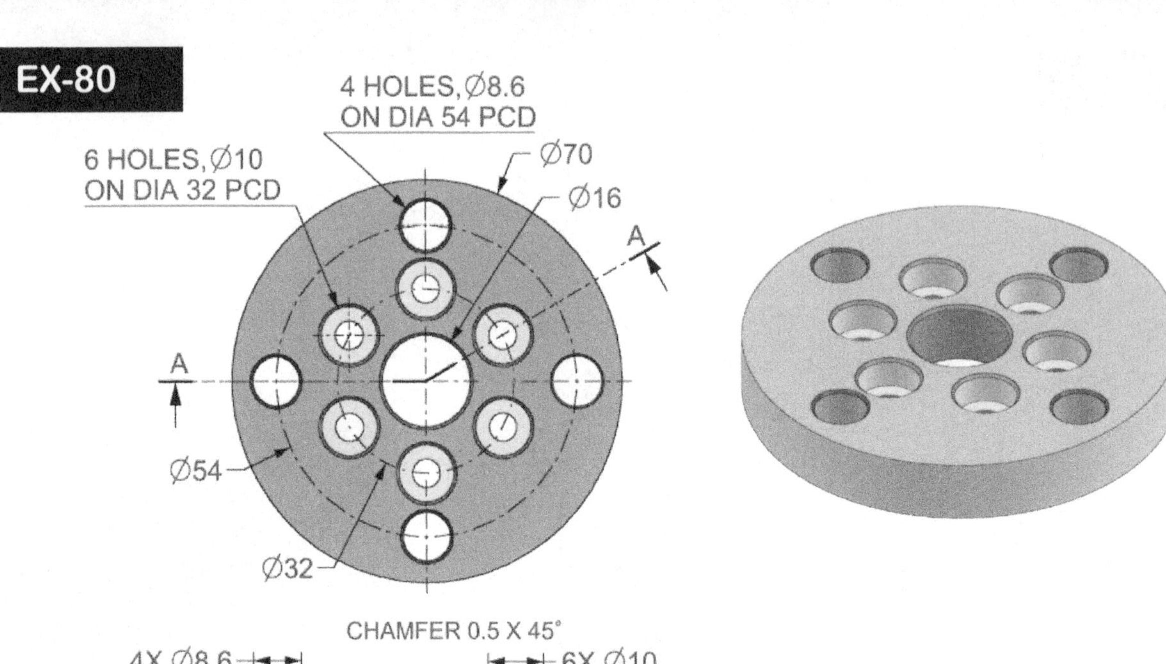

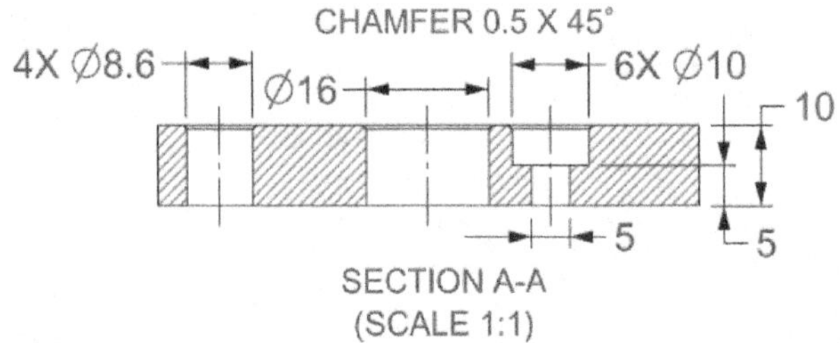

SECTION A-A
(SCALE 1:1)

EX-81

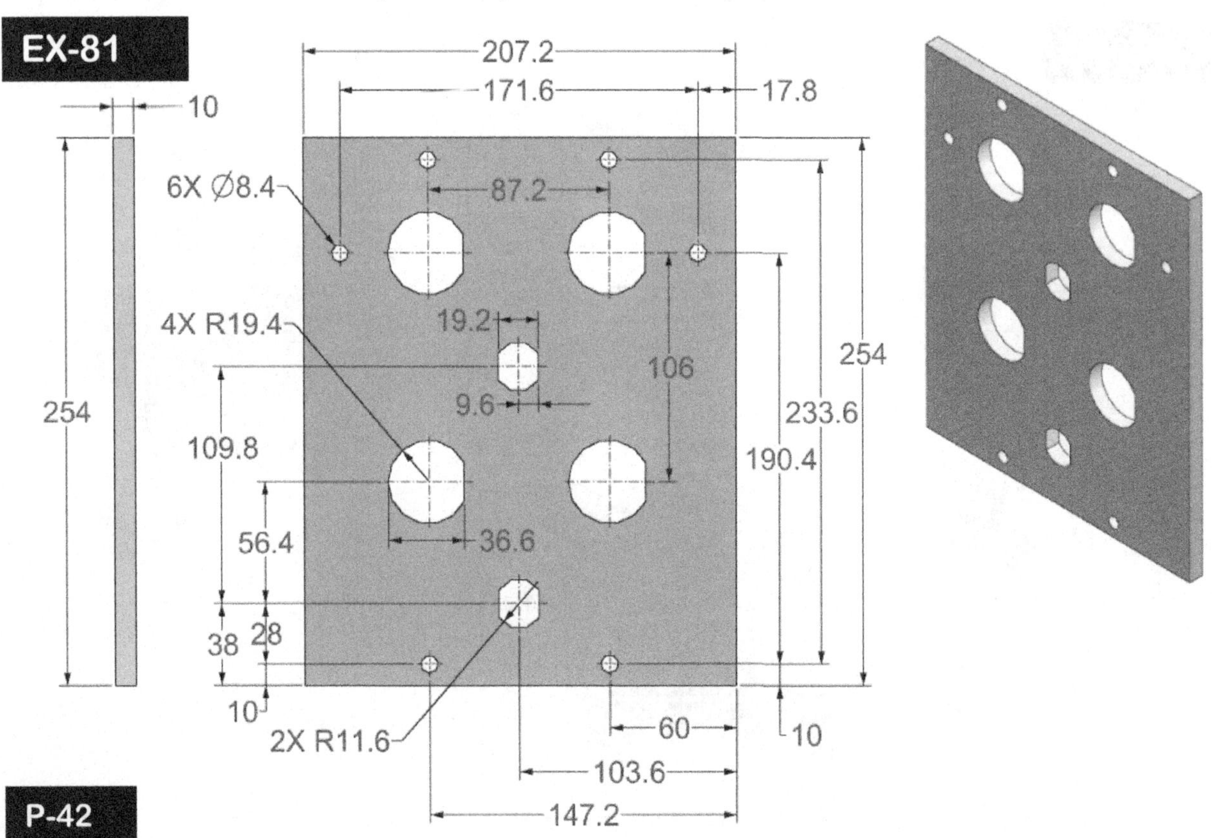

EX-82

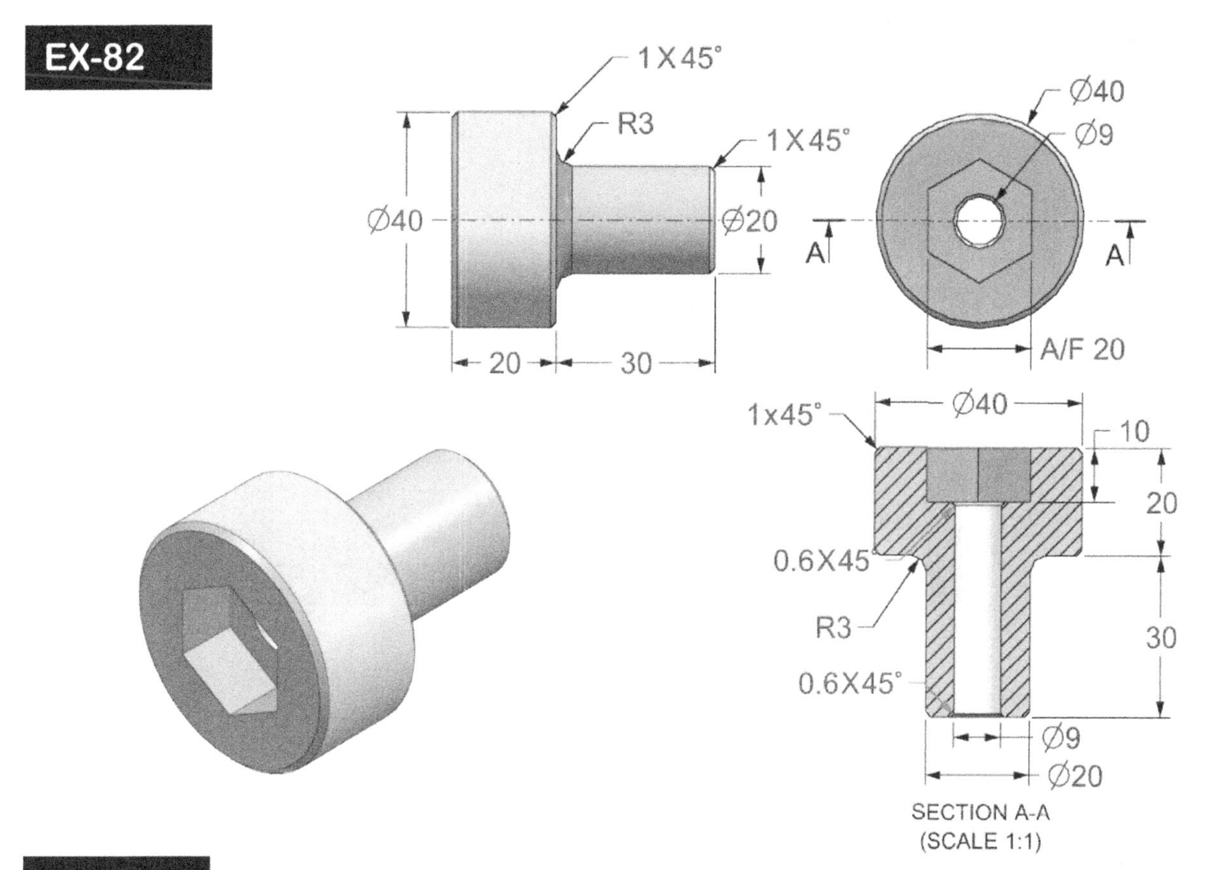
1X45°
R3
1X45°
Ø40
Ø20
20
30
Ø40
Ø9
A
A
A/F 20
1x45°
Ø40
10
20
0.6X45°
R3
30
0.6X45°
Ø9
Ø20
SECTION A-A
(SCALE 1:1)

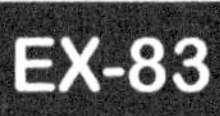

EX-83

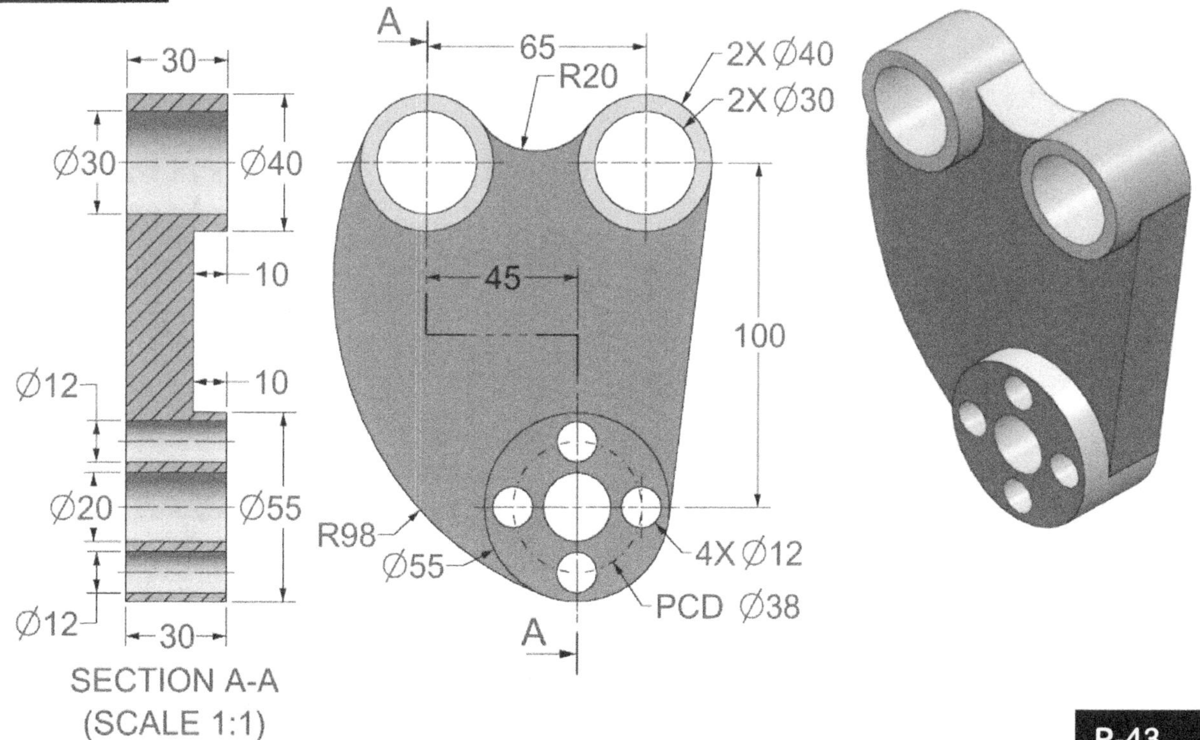
30
Ø30
Ø40
10
10
Ø12
Ø20
Ø55
Ø12
30
SECTION A-A
(SCALE 1:1)
A
65
R20
2X Ø40
2X Ø30
45
100
R98
Ø55
4X Ø12
PCD Ø38
A

EX-84

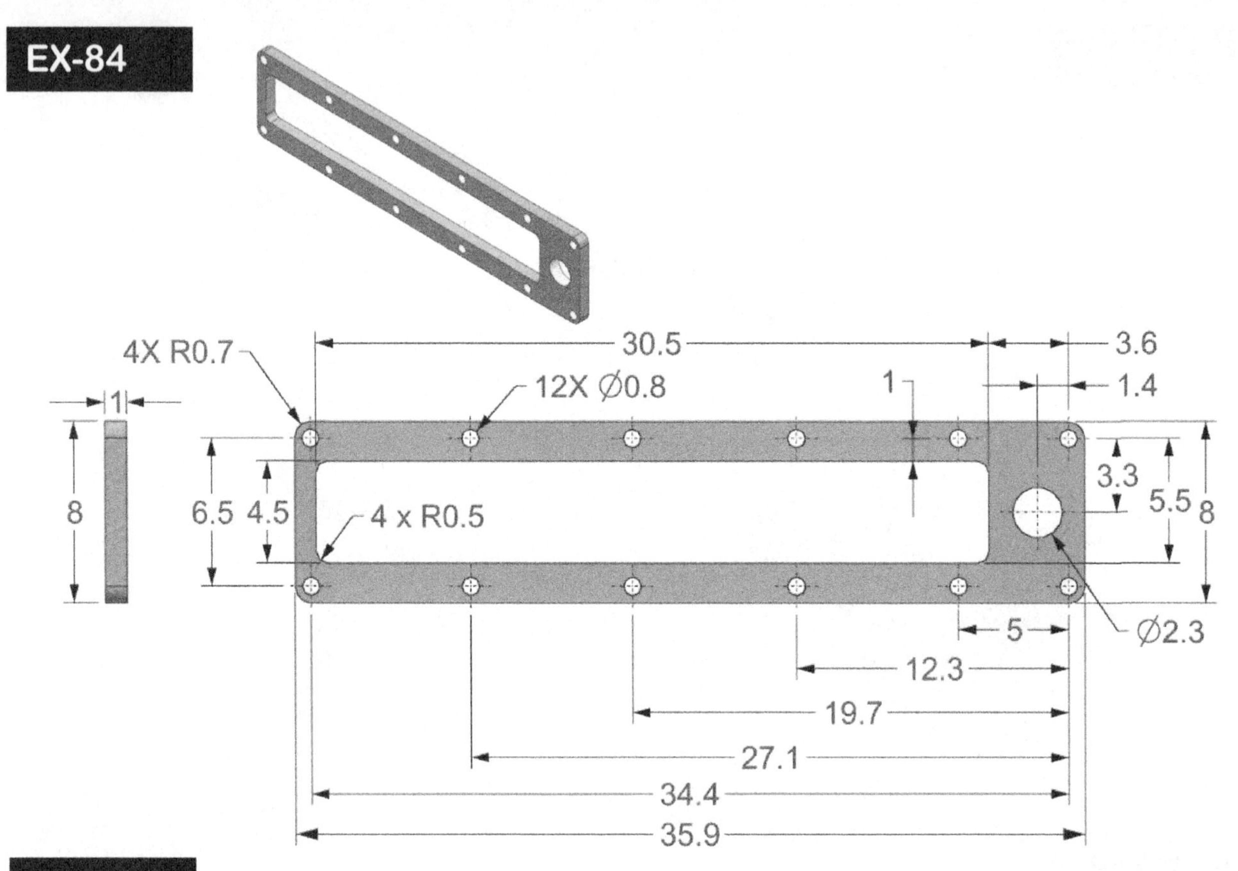
4X R0.7
30.5
3.6
12X Ø0.8
1
1.4
1
8
6.5
4.5
4 x R0.5
3.3
5.5
8
5
Ø2.3
12.3
19.7
27.1
34.4
35.9

EX-85

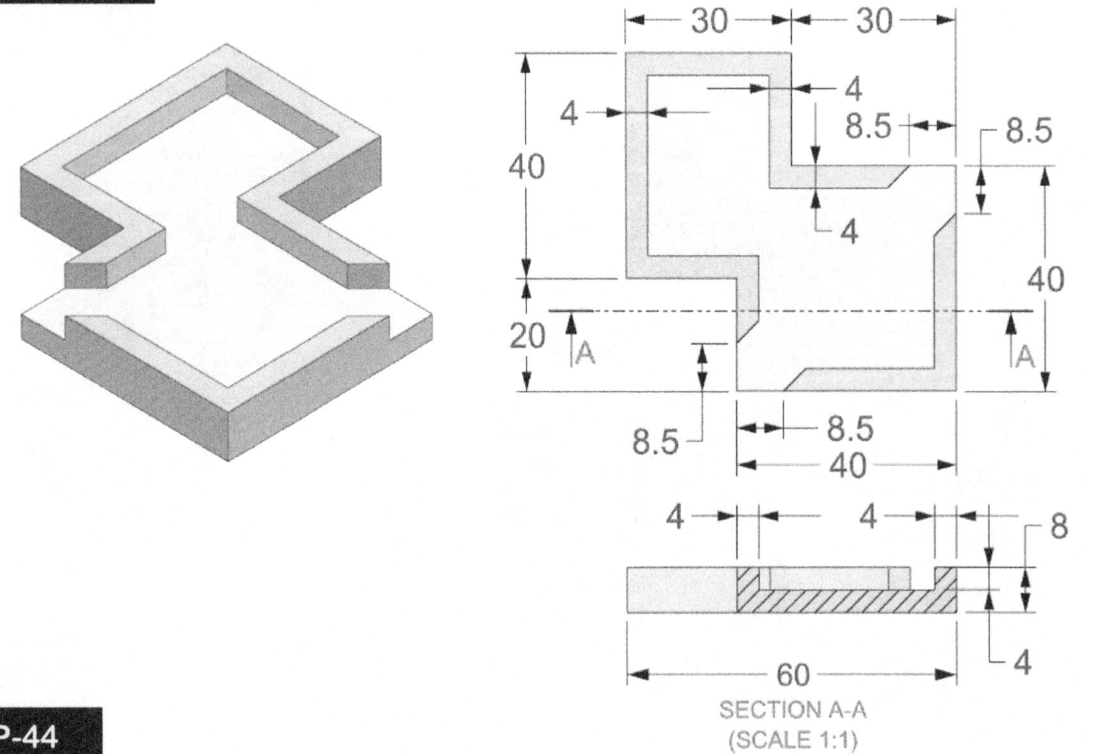
30
30
4
4
8.5
8.5
40
4
40
20
A
A
8.5
8.5
40
4
4
8
4
60
SECTION A-A
(SCALE 1:1)

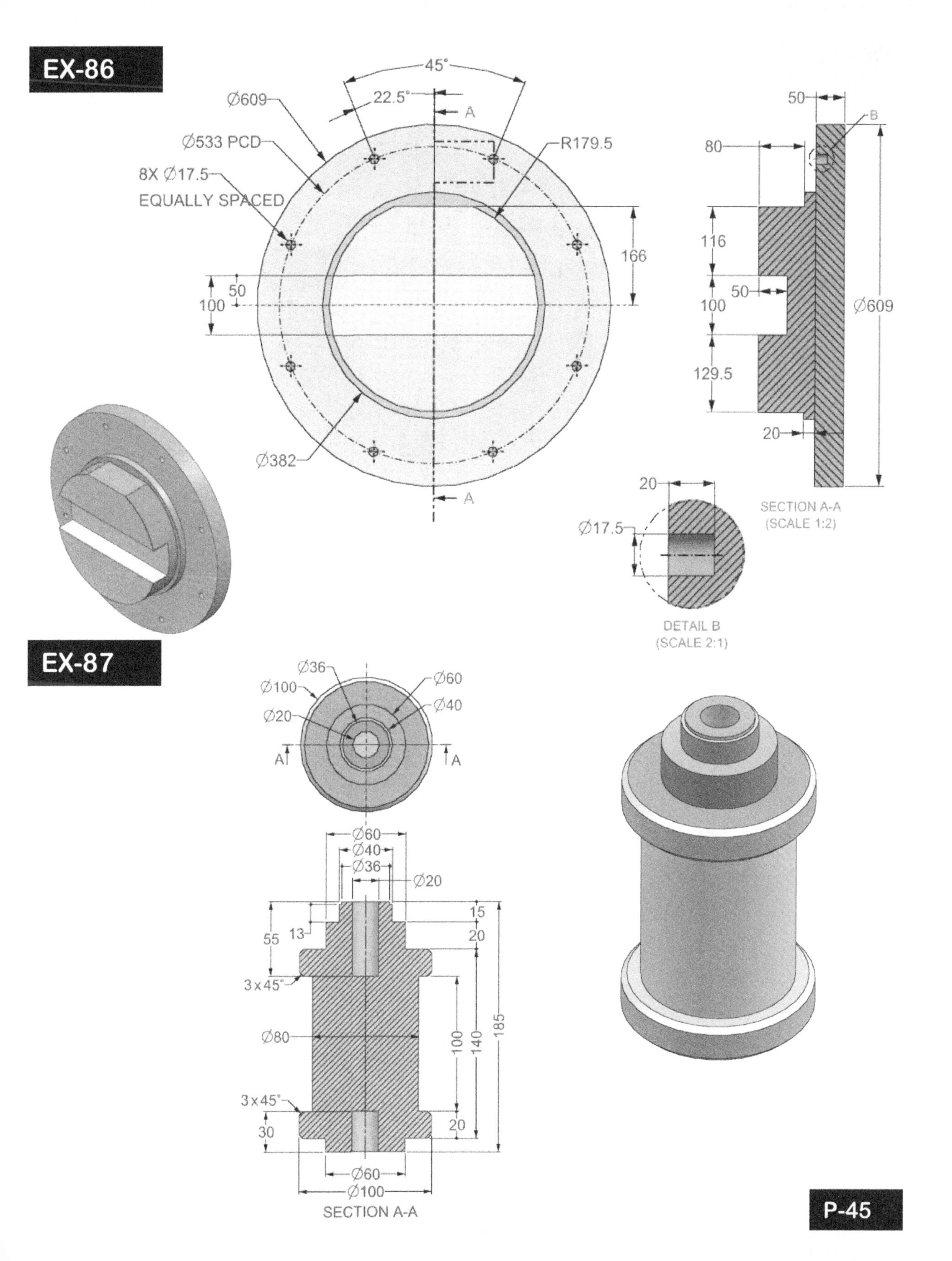
EX-86
45°
22.5°
Ø609
Ø533 PCD
8X Ø17.5
EQUALLY SPACED
R179.5
A
166
50
100
Ø382
50
80
B
116
50
100
Ø609
129.5
20
20
Ø17.5
DETAIL B
(SCALE 2:1)
SECTION A-A
(SCALE 1:2)
EX-87
Ø36
Ø100
Ø60
Ø40
Ø20
A
A
Ø60
Ø40
Ø36
Ø20
15
20
55
13
20
3 x 45°
Ø80
100
140
185
3 x 45°
30
20
Ø60
Ø100
SECTION A-A
P-45

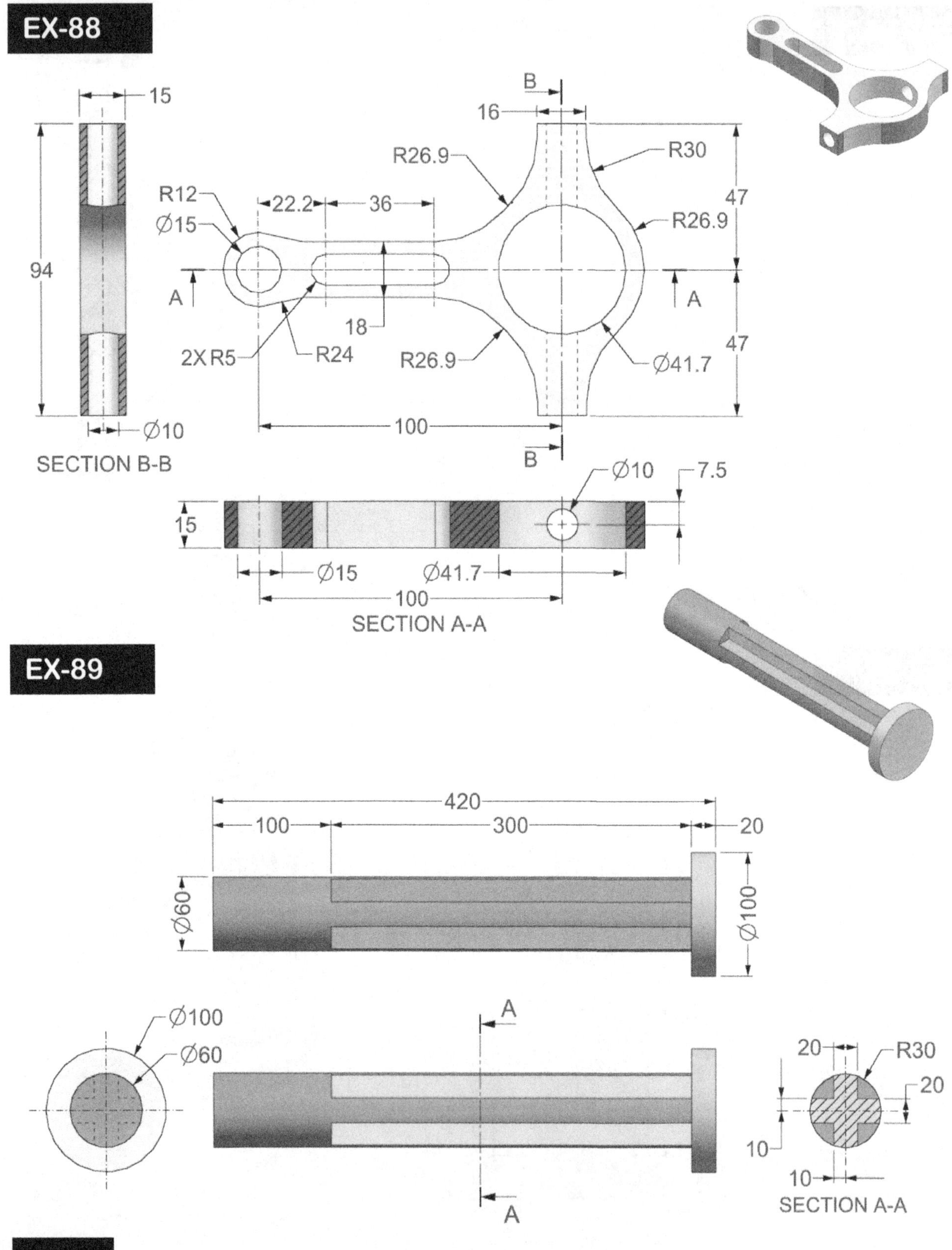
EX-88
15
94
Ø10
SECTION B-B
B
16
R26.9
R12
Ø15
22.2
36
R30
47
R26.9
A
A
18
2X R5
R24
R26.9
Ø41.7
47
100
B
Ø10
7.5
15
Ø15
Ø41.7
100
SECTION A-A
EX-89
420
100
300
20
Ø60
Ø100
Ø100
Ø60
A
A
20
R30
20
10
10
SECTION A-A

EX-90

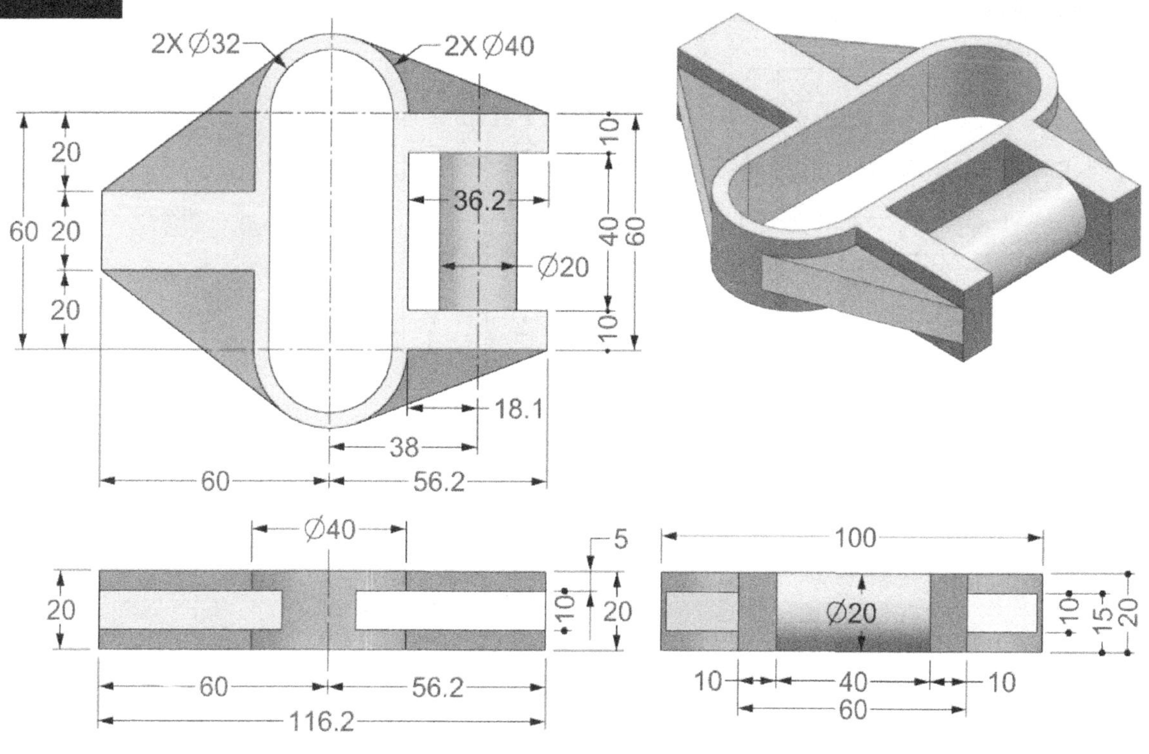

2X Ø32
2X Ø40
20
20
20
60
10
36.2
40
60
Ø20
10
18.1
38
60
56.2
Ø40
5
20
10
20
60
56.2
116.2
100
Ø20
10
15
20
10
40
10
60

EX-91

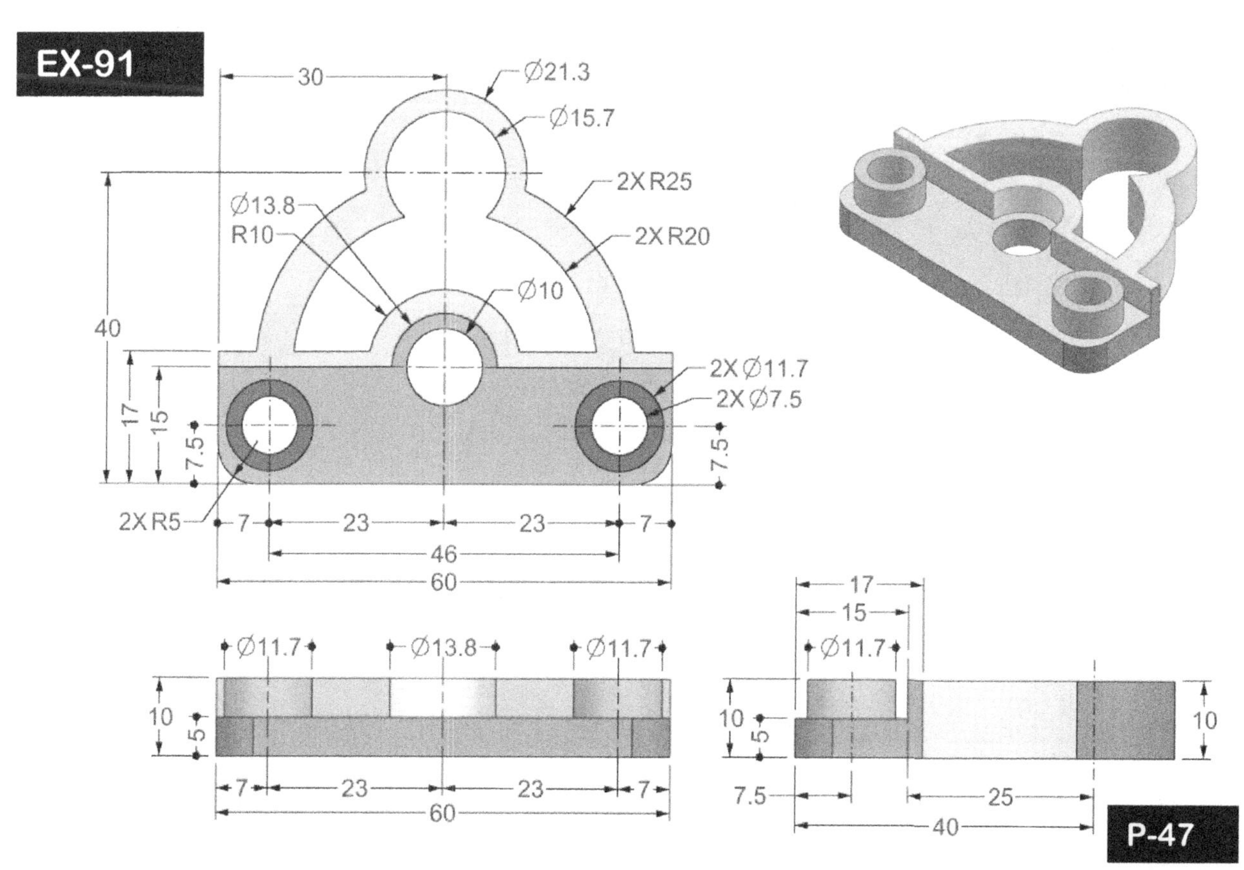

30
Ø21.3
Ø15.7
2X R25
Ø13.8
R10
2X R20
Ø10
40
17
15
7.5
2X Ø11.7
2X Ø7.5
7.5
2X R5
7
23
23
7
46
60
Ø11.7
Ø13.8
Ø11.7
10
5
7
23
23
7
60
17
15
Ø11.7
10
5
10
7.5
25
40

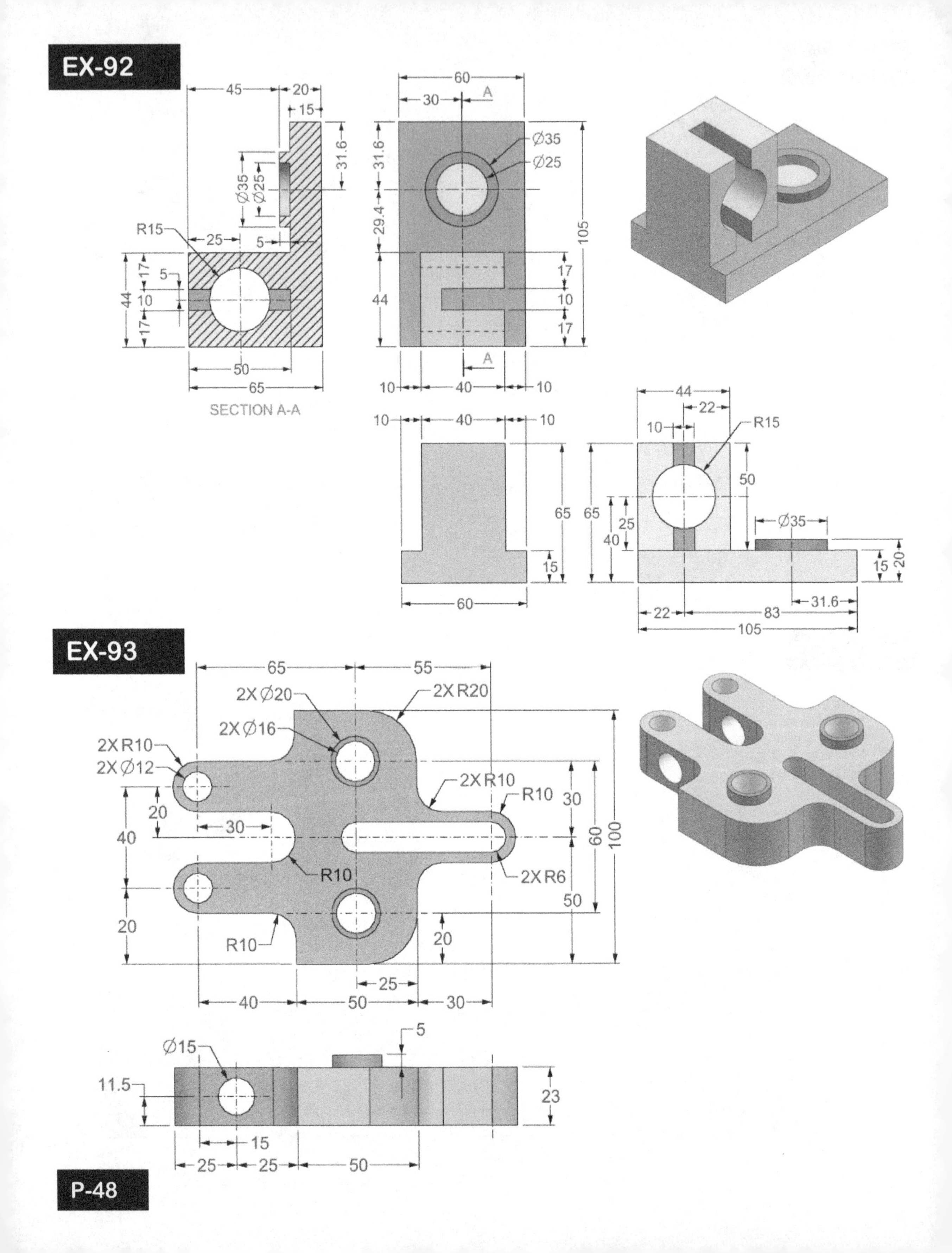
EX-92
45
20
15
31.6
Ø35
Ø25
R15
25
5
17
5
10
44
17
50
65
SECTION A-A
60
30
A
Ø35
Ø25
31.6
29.4
105
17
10
17
44
A
10
40
10
10
40
10
65
15
60
44
22
10
R15
50
65
65
25
40
Ø35
15
20
31.6
22
83
105
EX-93
65
55
2X Ø20
2X R20
2X Ø16
2X R10
2X Ø12
2X R10
R10
30
20
30
40
60
100
R10
2X R6
50
20
20
R10
25
40
50
30
Ø15
5
11.5
23
15
25
25
50
P-48

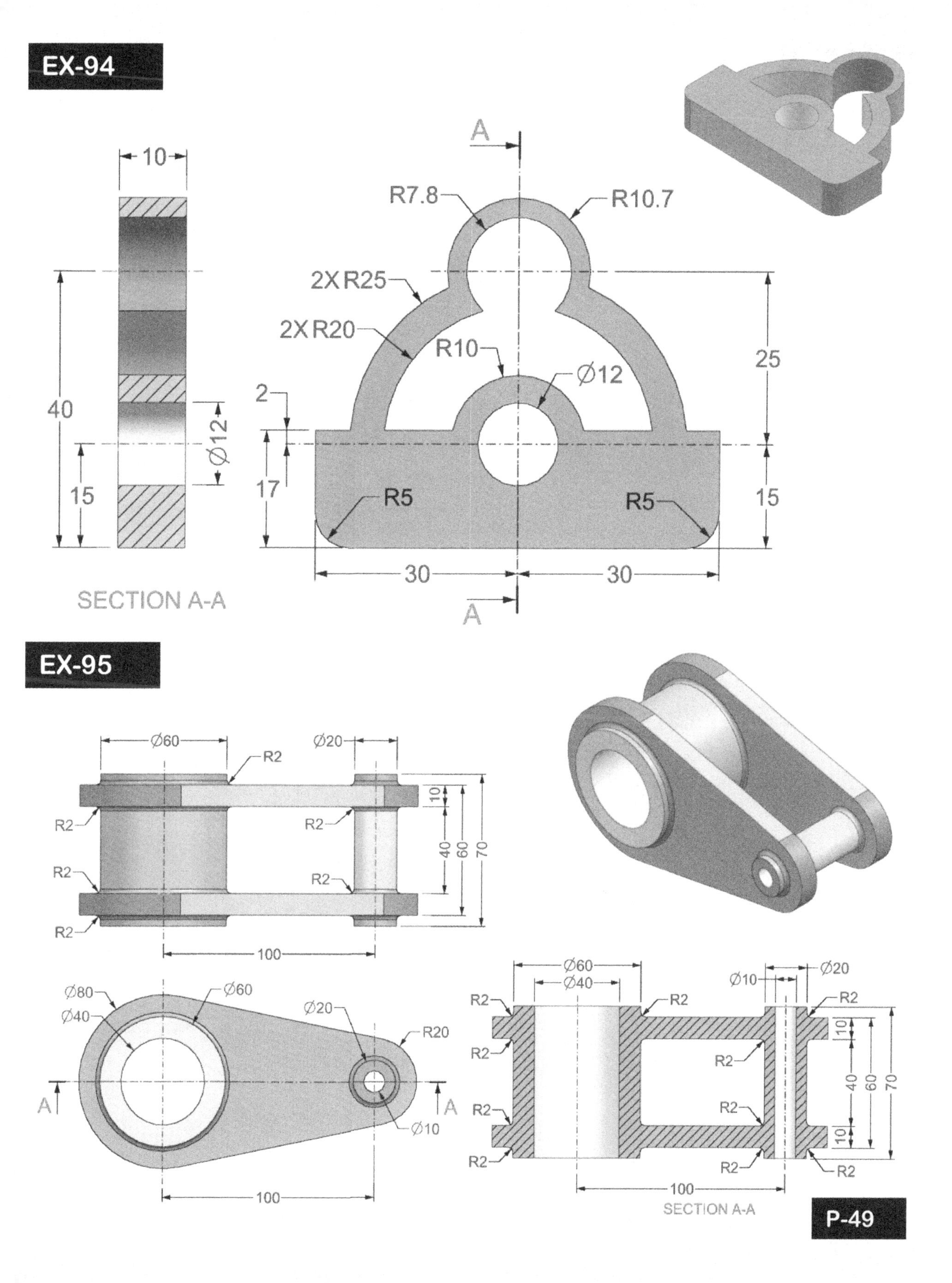
EX-94
SECTION A-A
10
40
15
Ø12
A
R7.8
R10.7
2X R25
2X R20
R10
Ø12
2
17
25
15
R5
R5
30
30
EX-95
Ø60
Ø20
R2
10
40
60
70
100
Ø80
Ø40
Ø10
R20
A
Ø60
Ø40
Ø10
Ø20
SECTION A-A
P-49

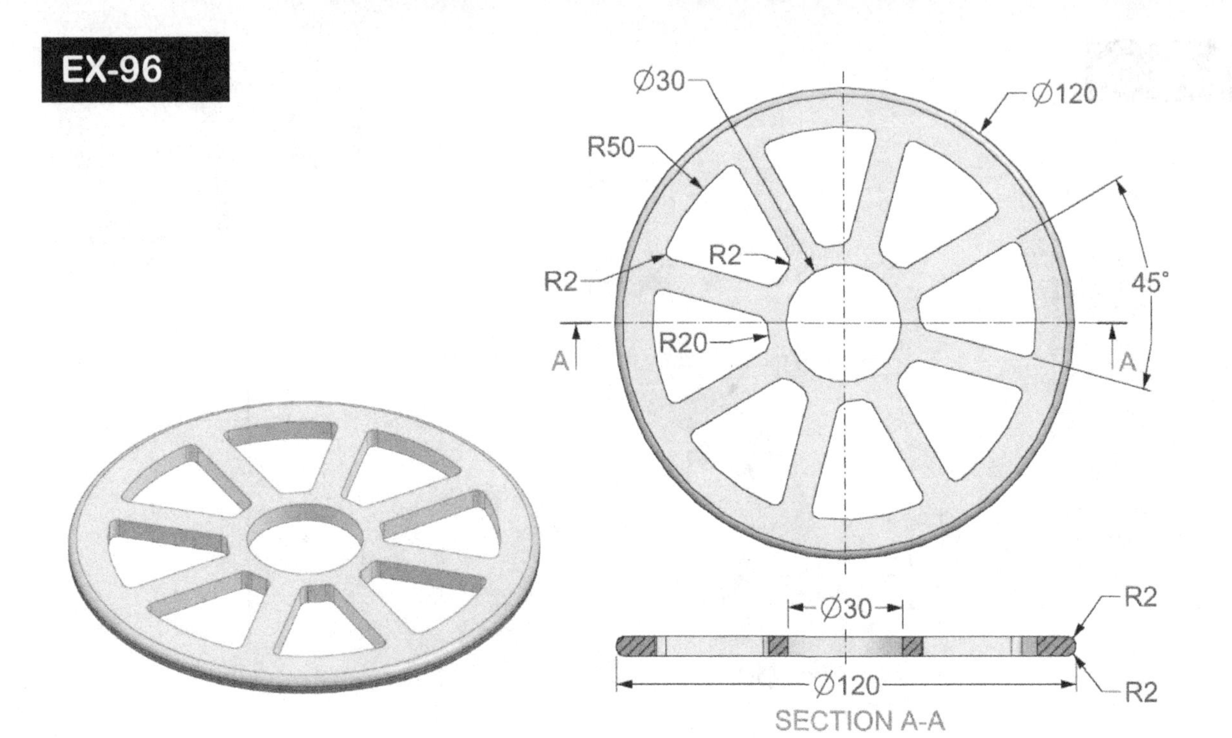
EX-96
Ø30
Ø120
R50
R2
R2
45°
A
A
R20
Ø30
R2
Ø120
R2
SECTION A-A

EX-97

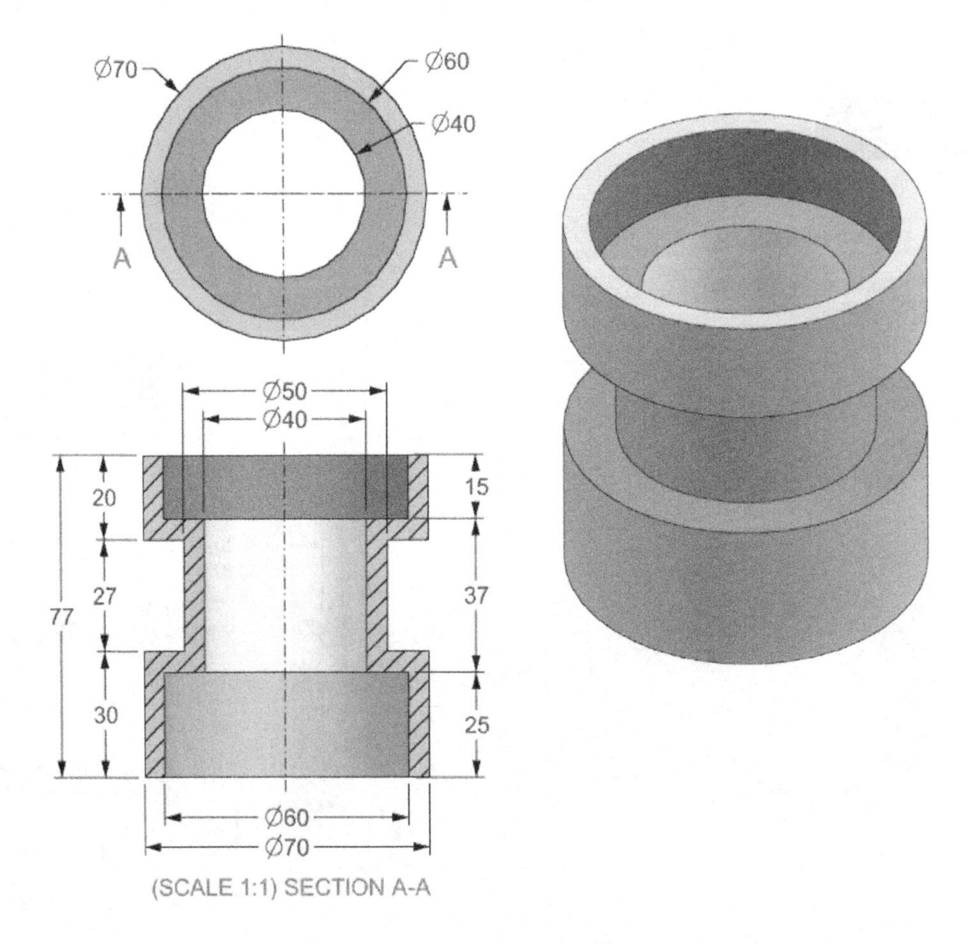
Ø70
Ø60
Ø40
A
A
Ø50
Ø40
20
15
27
77
37
30
25
Ø60
Ø70
(SCALE 1:1) SECTION A-A

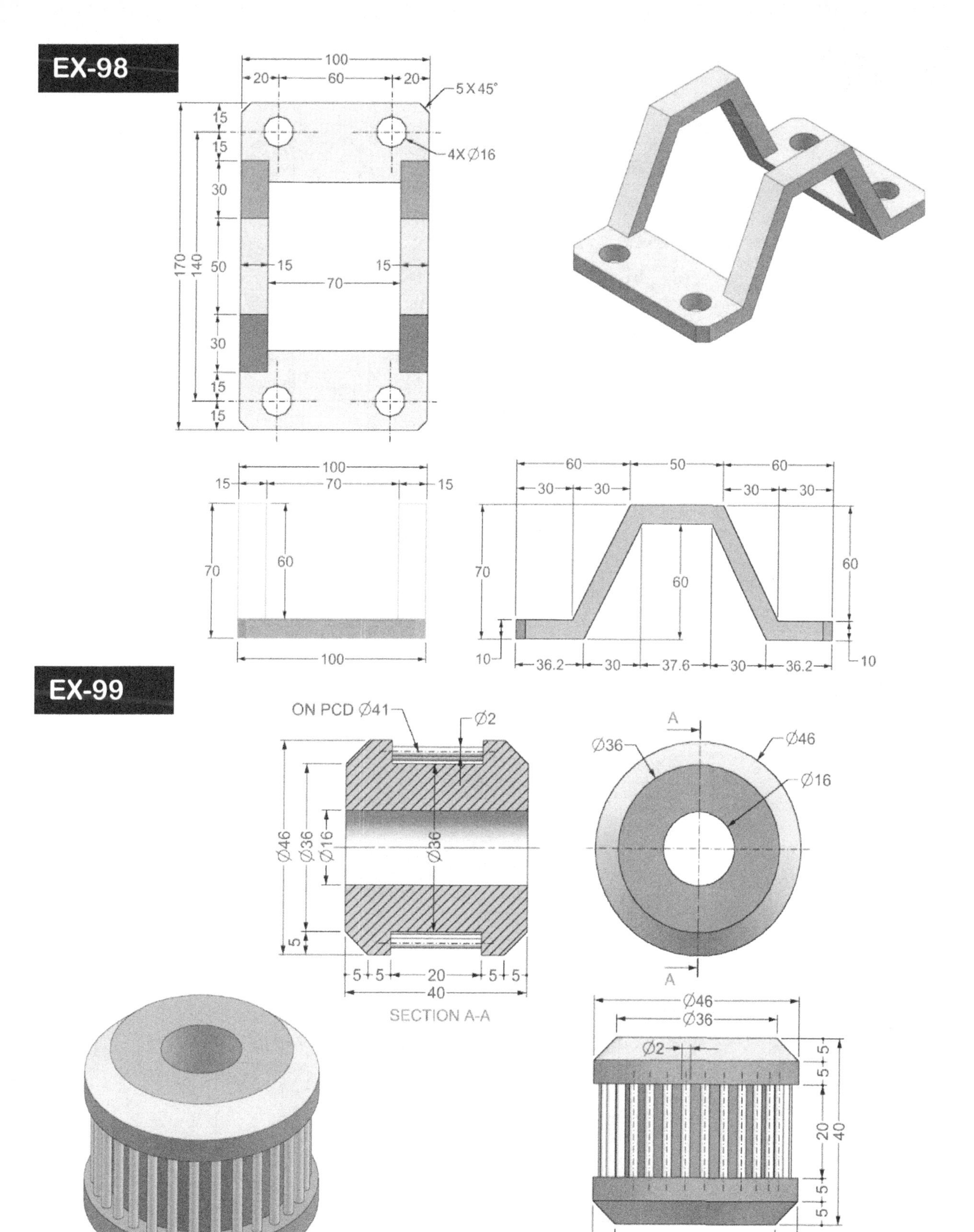
EX-98
100
20
60
20
5 X 45°
4X Ø16
15
15
30
50
30
15
15
170
140
15
15
70
100
15
70
15
70
60
100
60
50
60
30
30
30
30
70
60
60
10
36.2
30
37.6
30
36.2
10
EX-99
ON PCD Ø41
Ø2
Ø46
Ø36
Ø16
Ø36
5
5
5
20
5
5
40
SECTION A-A
A
Ø36
Ø46
Ø16
A
Ø46
Ø36
Ø2
5
5
20
40
5
5
Ø36
Ø46

EX-100

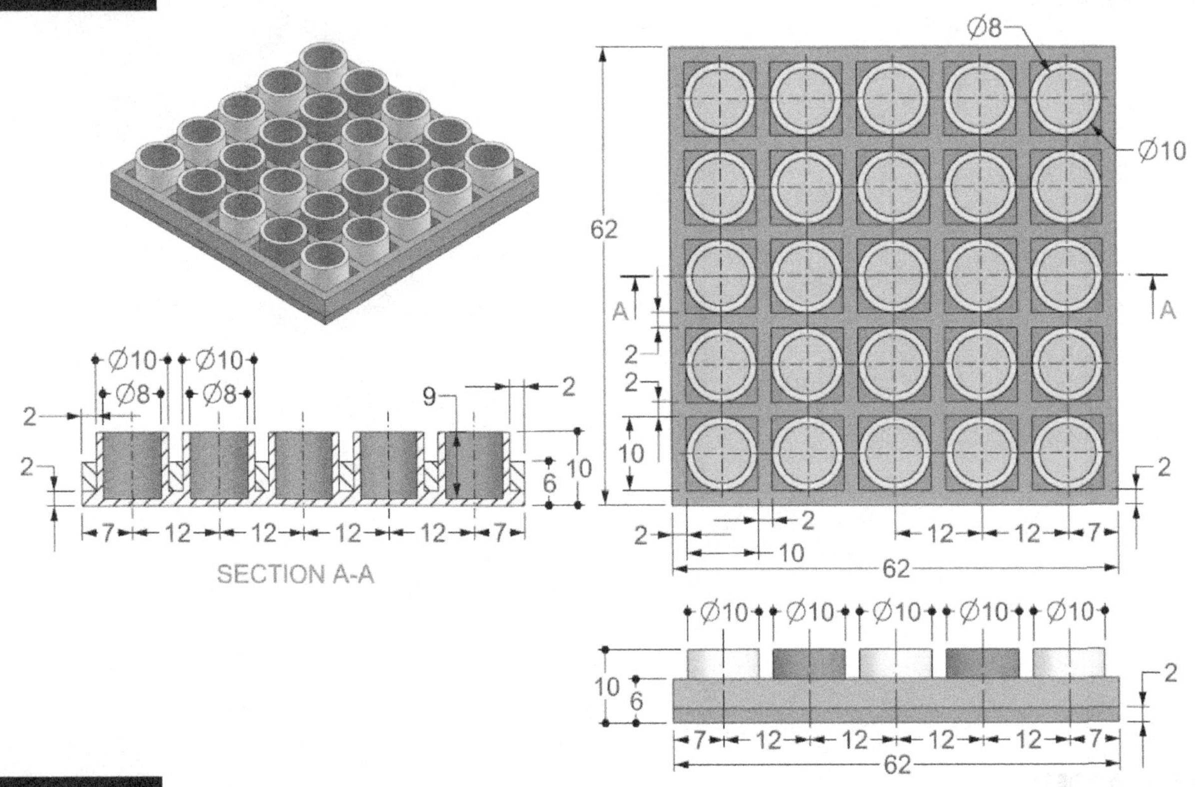

Ø8
Ø10
62
A
A
Ø10 Ø10
Ø8 Ø8
2
9
2
10
6
7 12 12 12 12 7
SECTION A-A
10
2
2
12 12 7
62
Ø10 Ø10 Ø10 Ø10 Ø10

EX-101

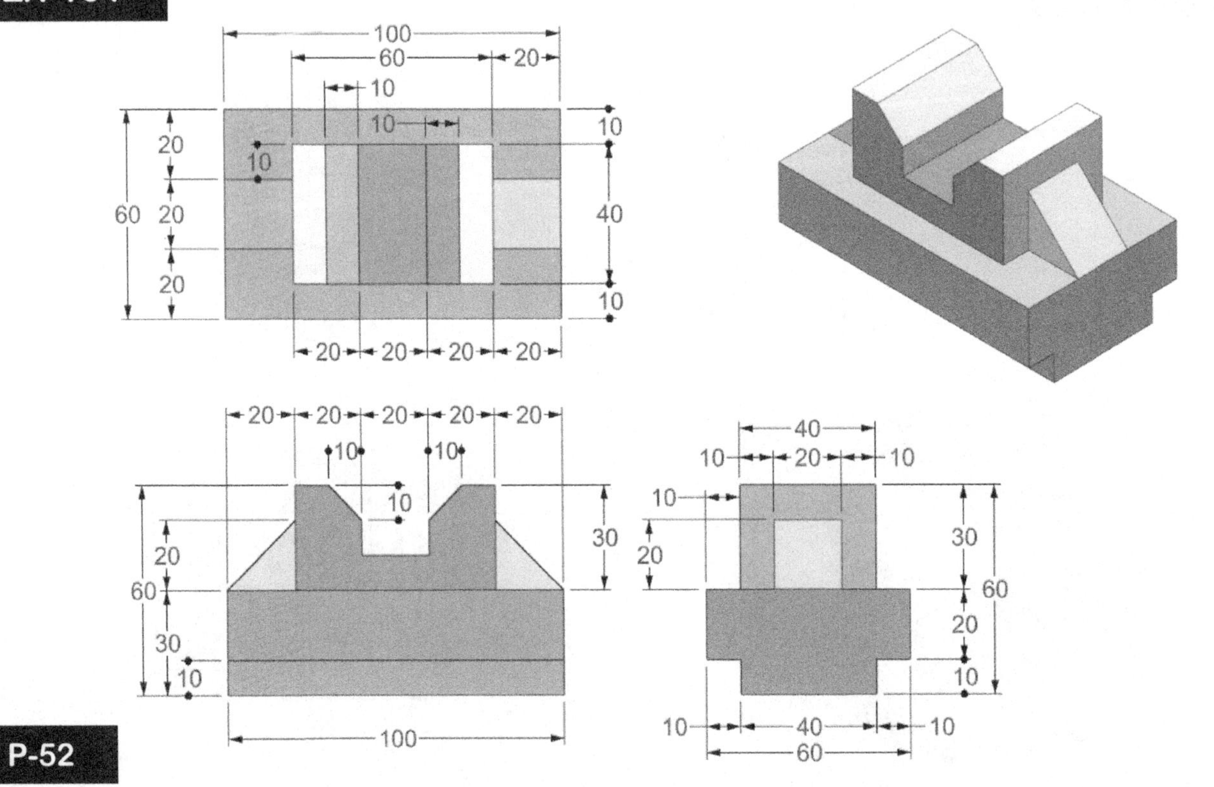

100
60
20
10
10
20
20
20
60
40
20 20 20 20
20 20 20 20 20
30
100
40
10 20 10

EX-102

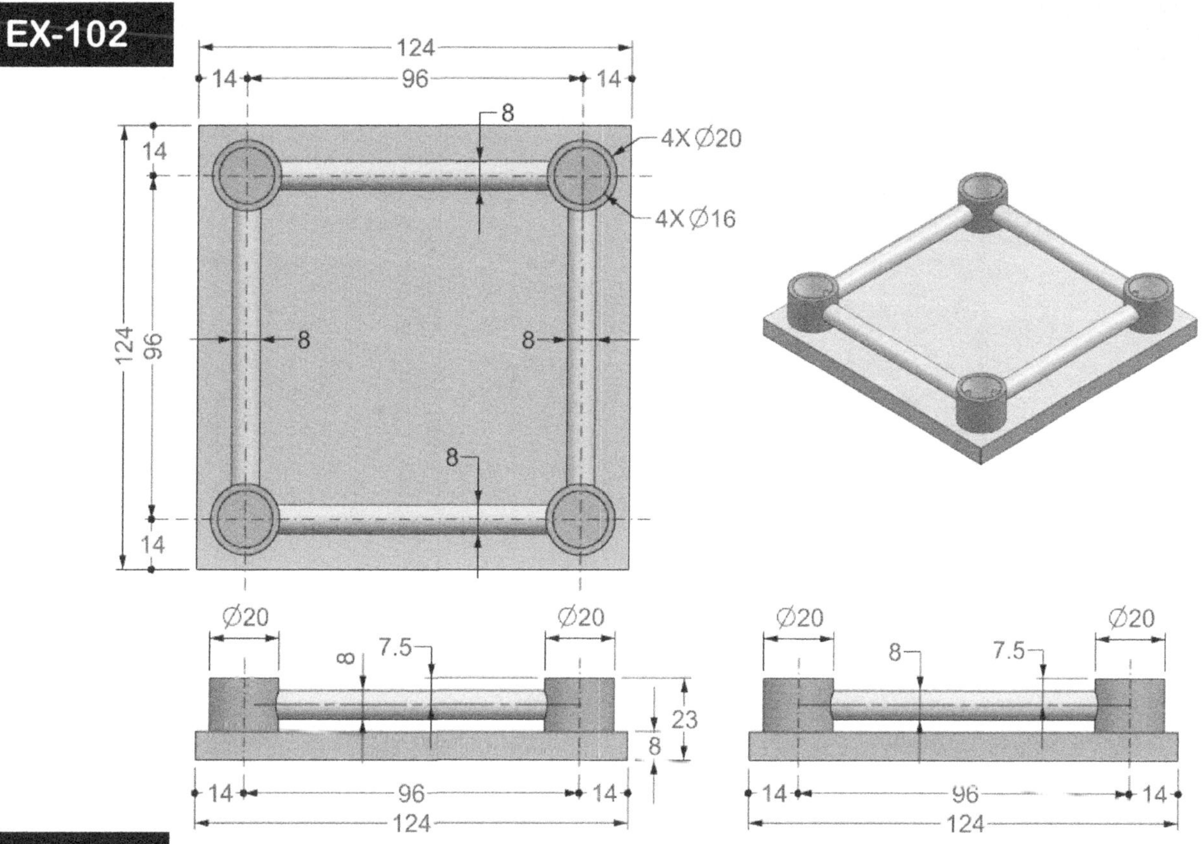
124
14
96
14
8
4X Ø20
4X Ø16
124
96
14
14
8
8
8
Ø20
Ø20
7.5
8
23
8
14
96
14
124
Ø20
Ø20
8
7.5
14
96
14
124

EX-103

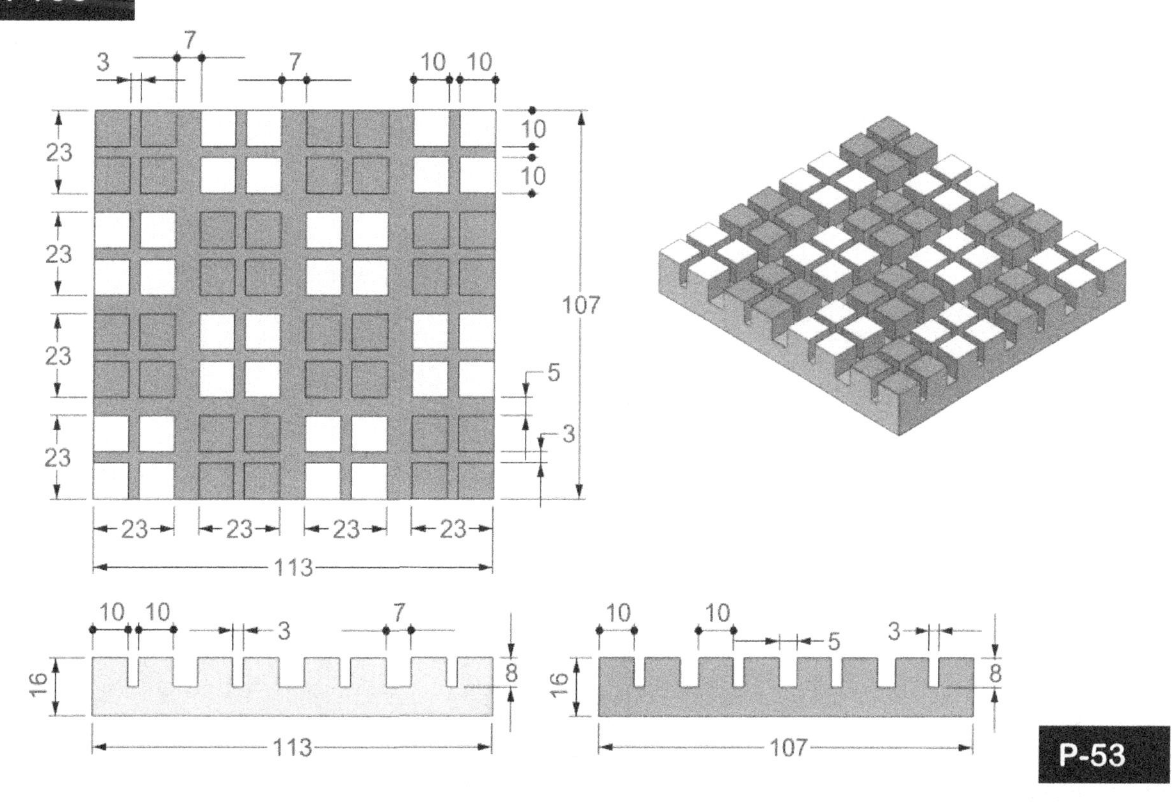
3
7
7
10
10
10
10
23
23
23
23
107
5
3
23
23
23
23
113
10
10
3
7
16
8
113
10
10
5
3
16
8
107

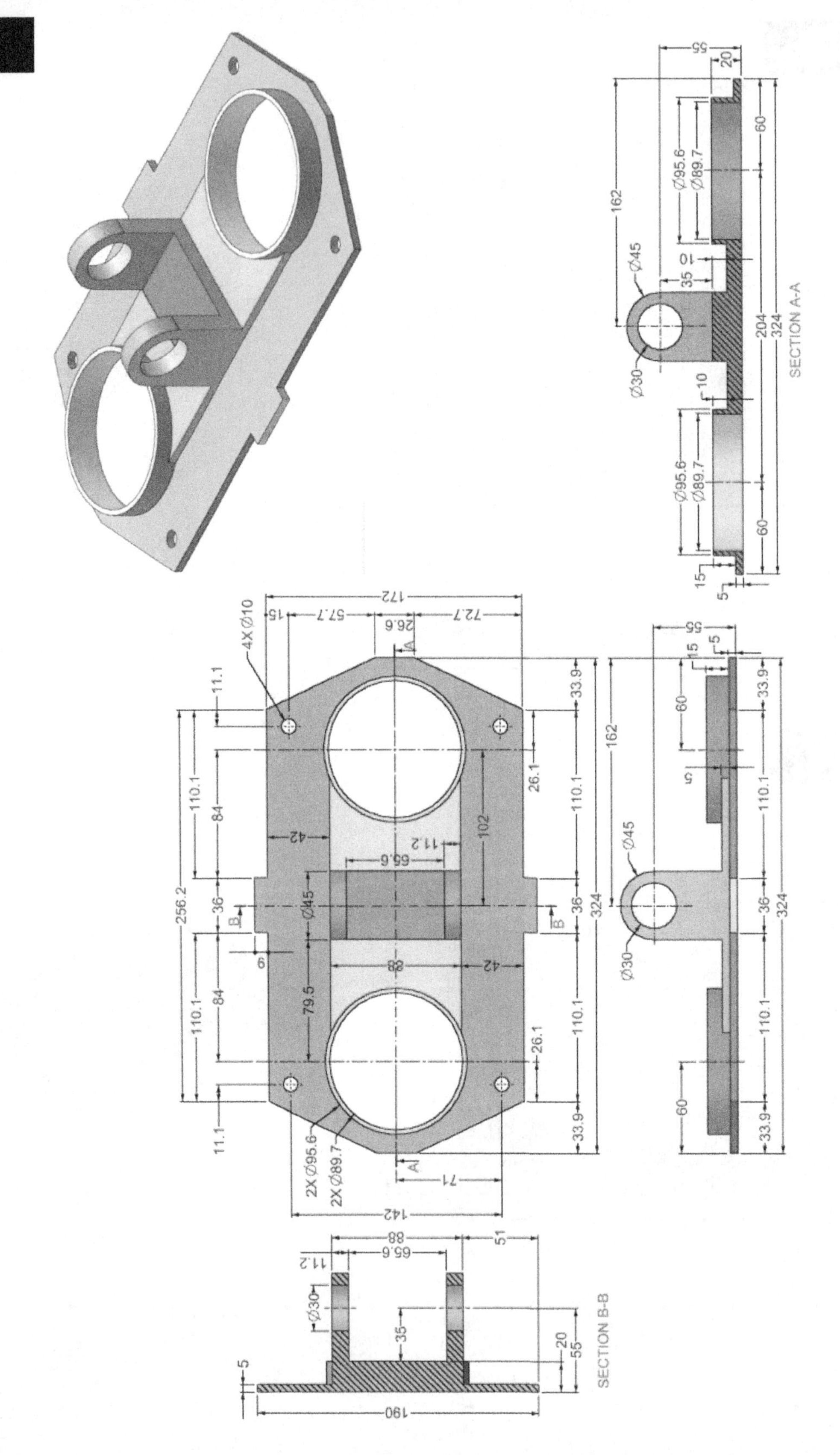
SECTION A-A
SECTION B-B

EX-105

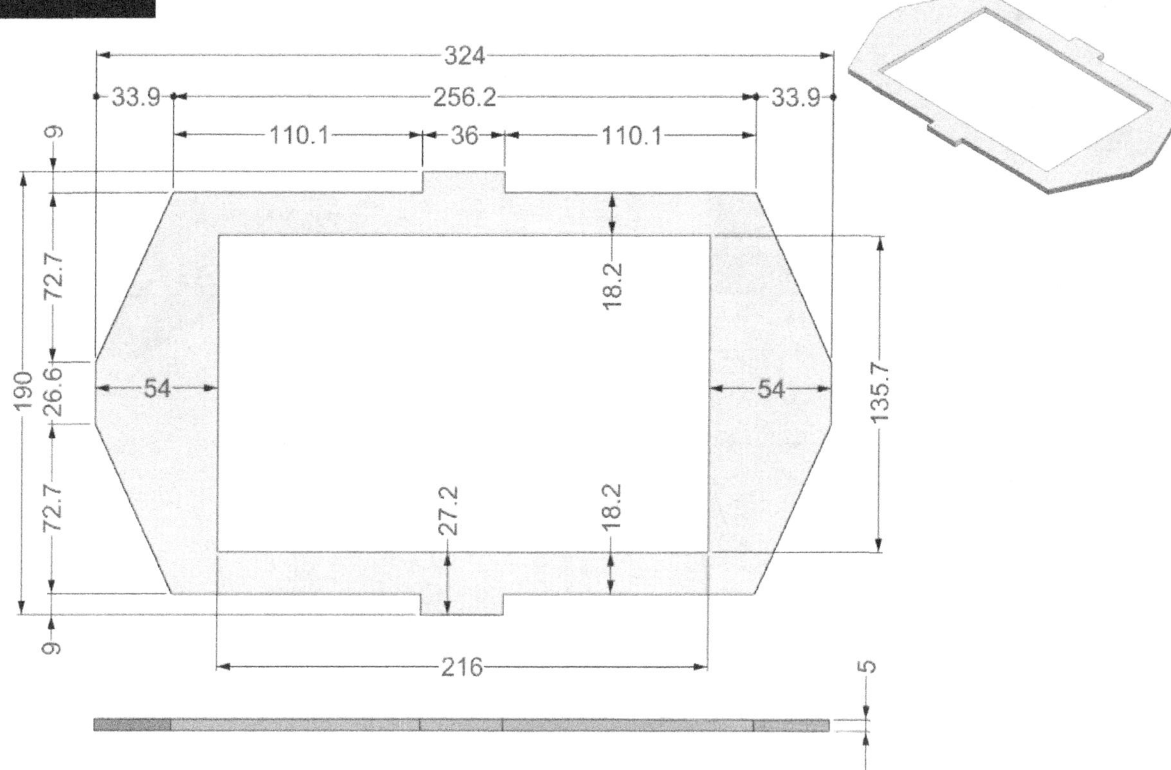

324
33.9
256.2
33.9
110.1
36
110.1
9
72.7
18.2
190
26.6
54
54
135.7
72.7
27.2
18.2
9
216
5

EX-106

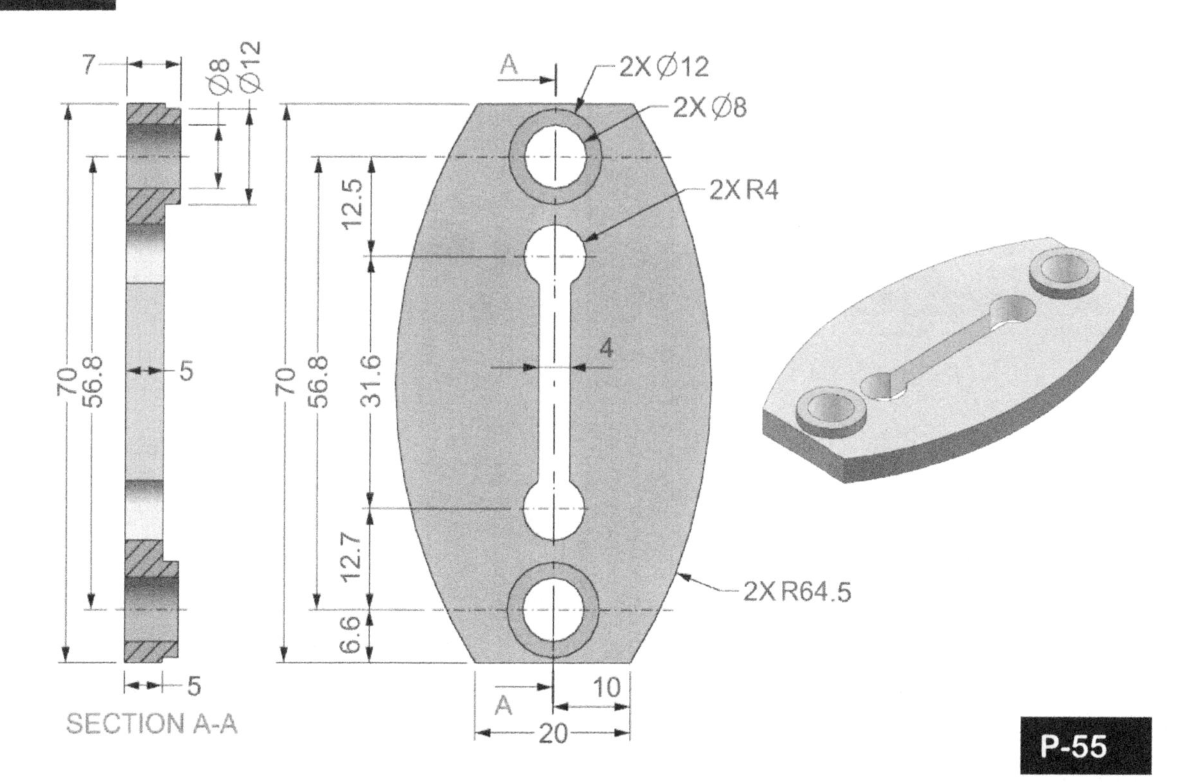

7
Ø8
Ø12
70
56.8
5
5
SECTION A-A
A
2X Ø12
2X Ø8
2X R4
70
56.8
12.5
31.6
4
12.7
6.6
2X R64.5
A
10
20

EX-107

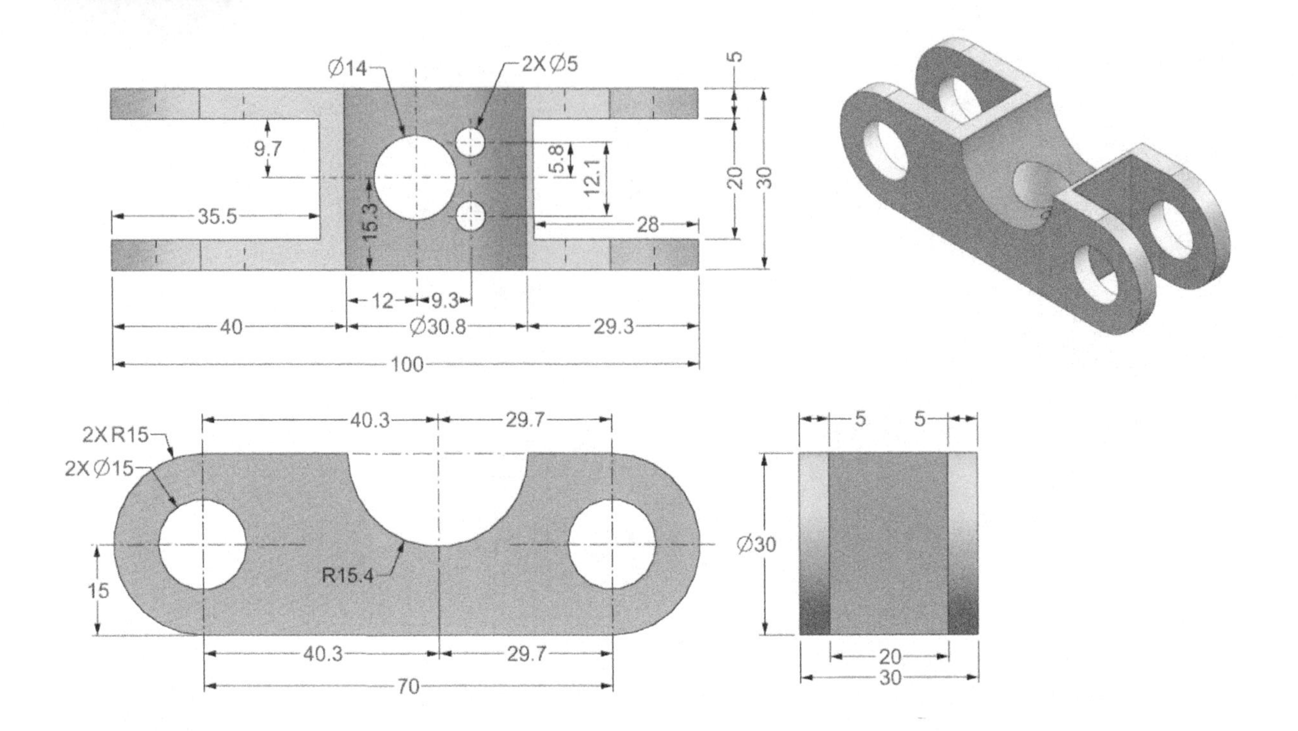

Ø14
2X Ø5
5
9.7
5.8
12.1
20
30
35.5
15.3
28
12
9.3
40
Ø30.8
29.3
100
40.3
29.7
2X R15
2X Ø15
R15.4
15
40.3
29.7
70
5
5
Ø30
20
30

EX-108

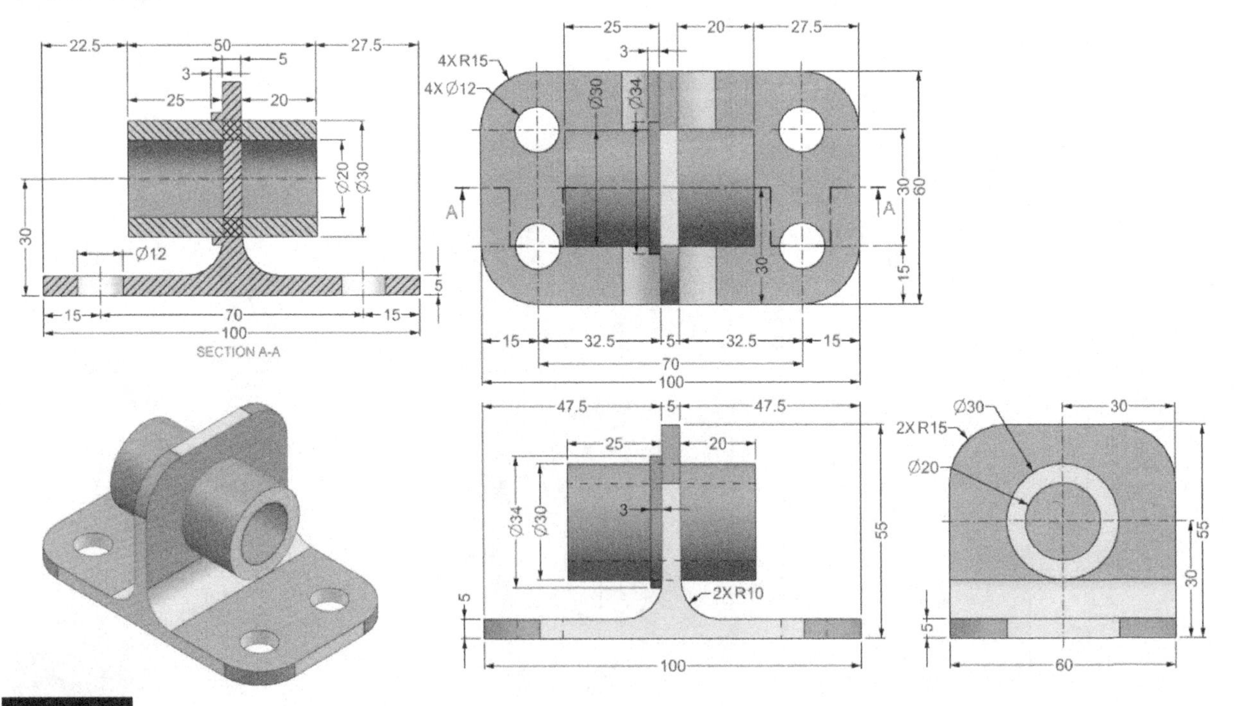

22.5
50
27.5
3
5
25
20
Ø20
Ø30
30
Ø12
5
15
70
15
100
SECTION A-A
25
20
27.5
3
4X R15
4X Ø12
Ø30
Ø34
A
A
30
60
30
15
15
32.5
5
32.5
15
70
100
47.5
5
47.5
25
20
Ø34
Ø30
3
55
2X R10
5
100
Ø30
30
2X R15
Ø20
55
30
5
60

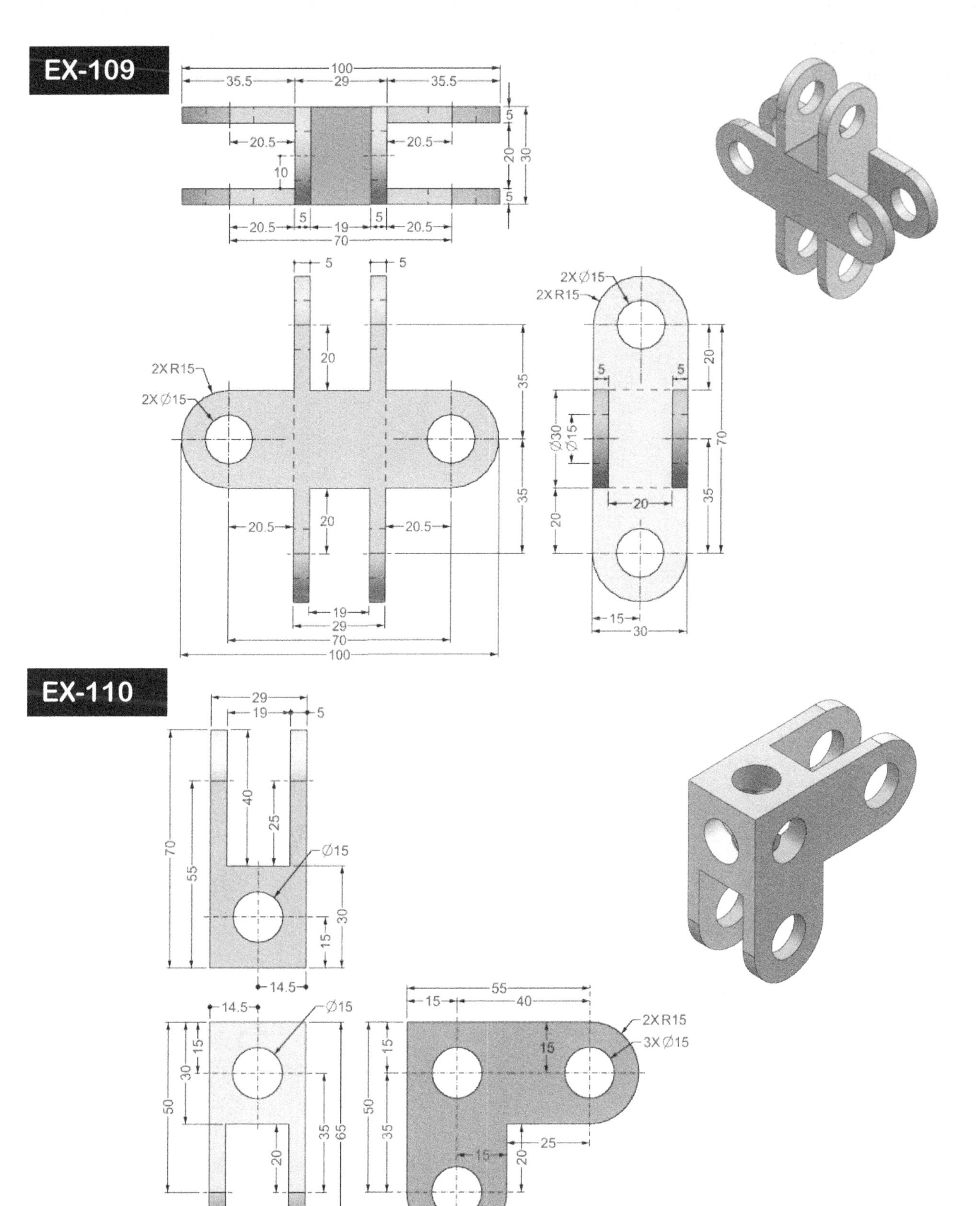
EX-109
100
35.5
29
35.5
5
20
30
5
20.5
20.5
10
20.5
5
19
5
20.5
70
5
5
2X Ø15
2X R15
2X R15
2X Ø15
20
35
35
20
20.5
20.5
19
29
70
100
20
5
5
Ø30
Ø15
70
35
20
20
15
30
EX-110
29
19
5
70
55
40
25
Ø15
30
15
14.5
14.5
Ø15
15
30
50
20
35
65
19
5
29
55
15
40
2X R15
3X Ø15
15
15
50
35
25
15
20

EX-111

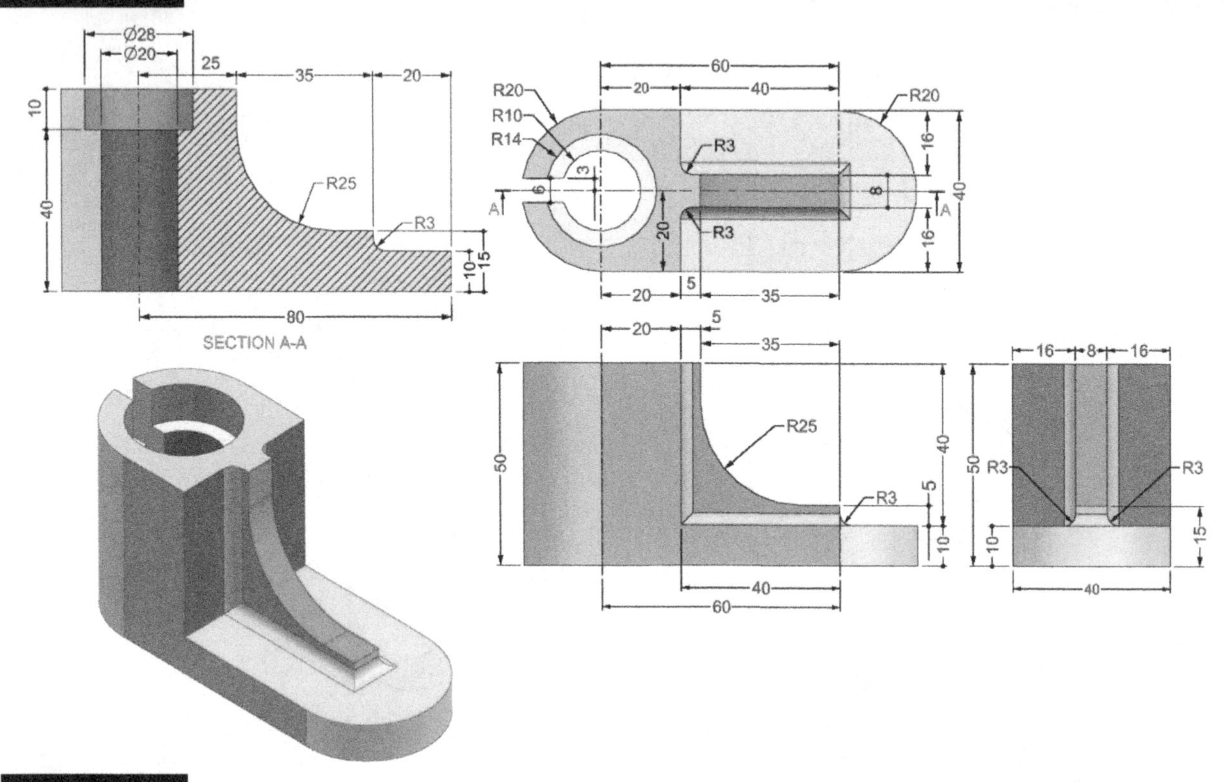
Ø28
Ø20
25
35
20
10
40
R25
R3
10
15
80
SECTION A-A
60
20
40
R20
R10
R14
R20
R3
3
6
8
16
40
A
A
20
R3
16
5
20
35
5
20
35
50
R25
40
R3
5
10
40
60
16
8
16
50
R3
R3
10
15
40

EX-112

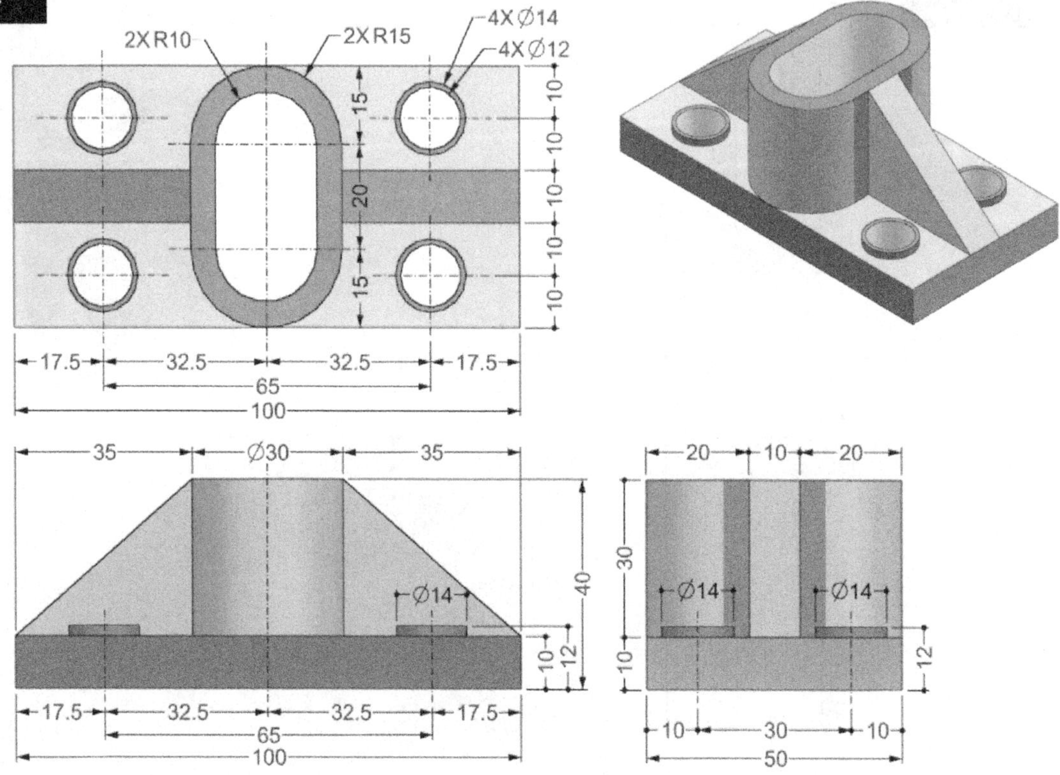
2X R10
2X R15
4X Ø14
4X Ø12
15
20
15
10
10
10
10
10
10
17.5
32.5
32.5
17.5
65
100
35
Ø30
35
Ø14
40
10
12
17.5
32.5
32.5
17.5
65
100
20
10
20
30
Ø14
Ø14
10
12
10
30
10
50

EX-113

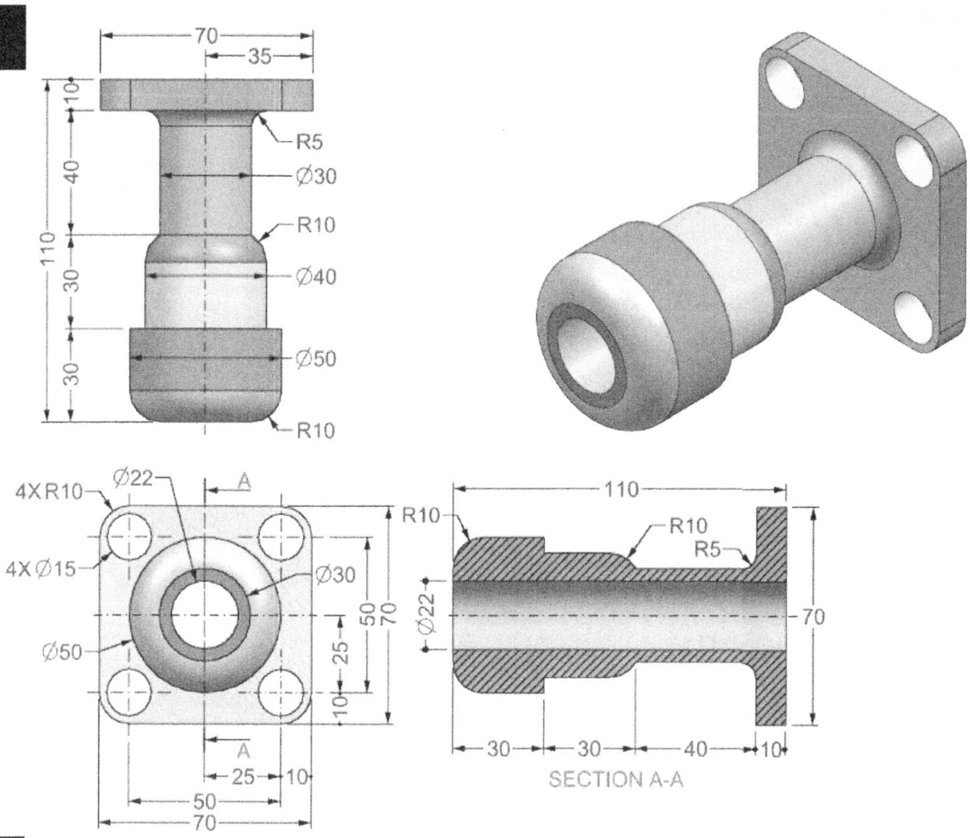
70
35
10
40
110
30
30
R5
Ø30
R10
Ø40
Ø50
R10
4X R10
Ø22
A
4X Ø15
Ø30
Ø50
50
70
25
10
A
25
10
50
70
110
R10
R10
R5
Ø22
70
30
30
40
10
SECTION A-A

EX-114

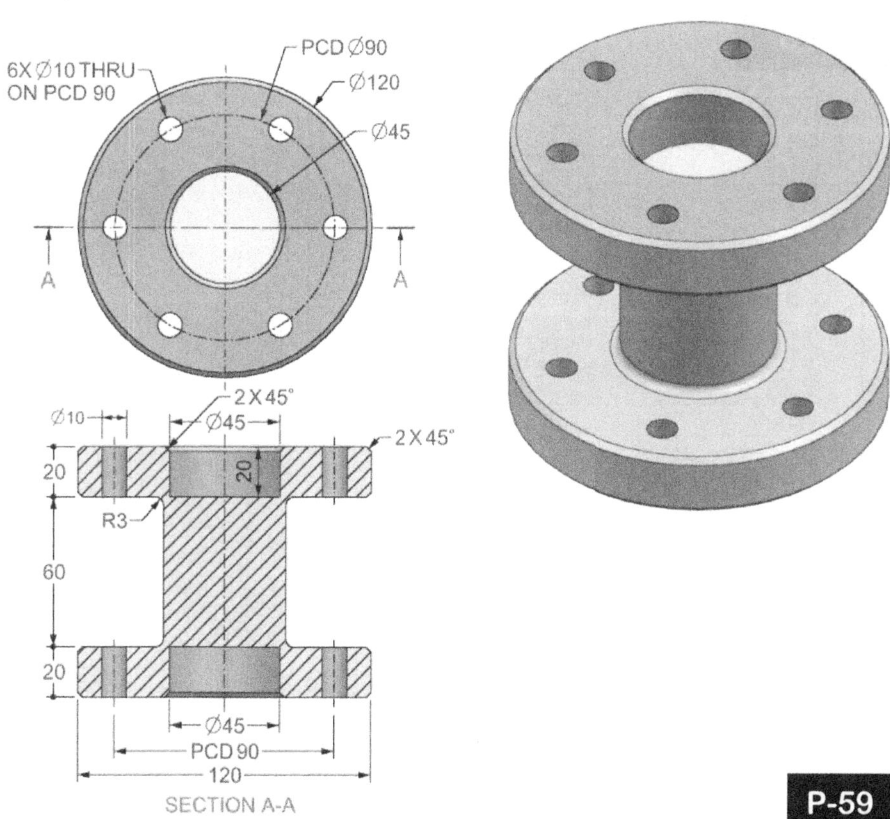
6X Ø10 THRU
ON PCD 90
PCD Ø90
Ø120
Ø45
A
A
2 X 45°
Ø10
Ø45
2 X 45°
20
20
R3
60
20
Ø45
PCD 90
120
SECTION A-A

EX-115

Ø120
6X Ø10
6X Ø8
PCD Ø90
Ø68
Ø45
A
A

Ø120
Ø68
Ø10
R2
10
10
20
120
60
20
10
PCD 90

Ø120
PCD 90
Ø68
Ø45
Ø10
2 X 45°
40
20
60
20
R3
R3
Ø8
Ø55
Ø50
20
Ø45
Ø8
SECTION A-A

EX-116

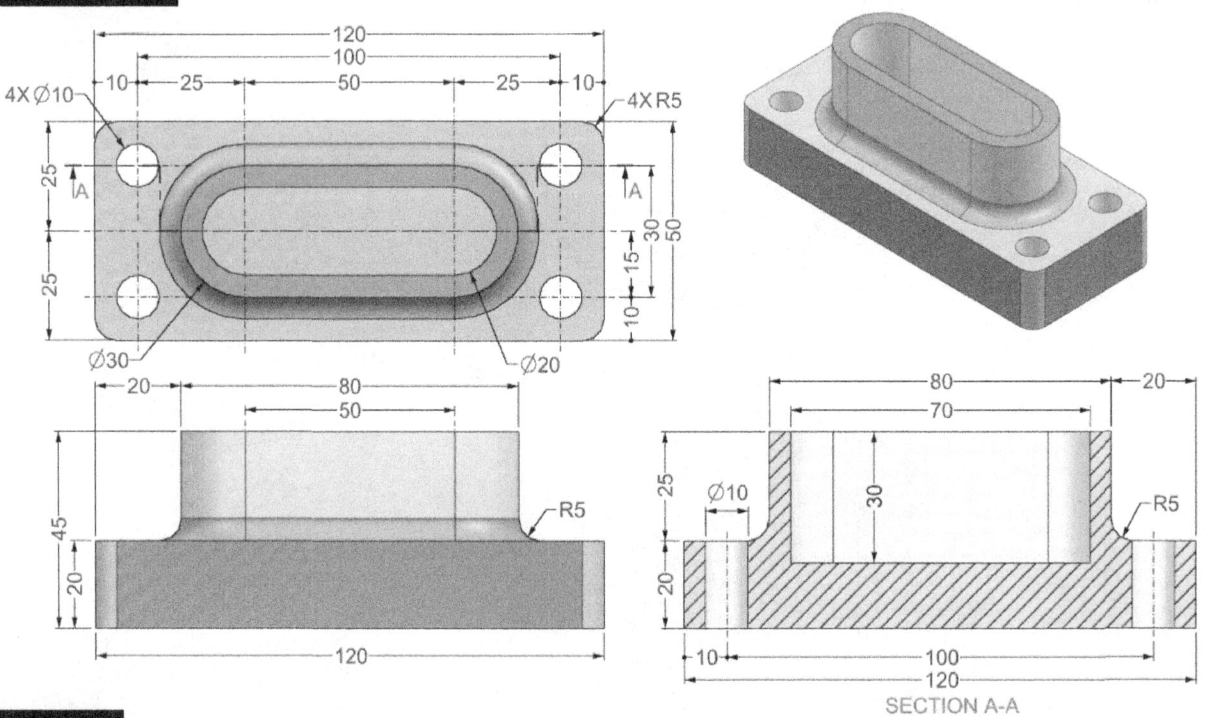

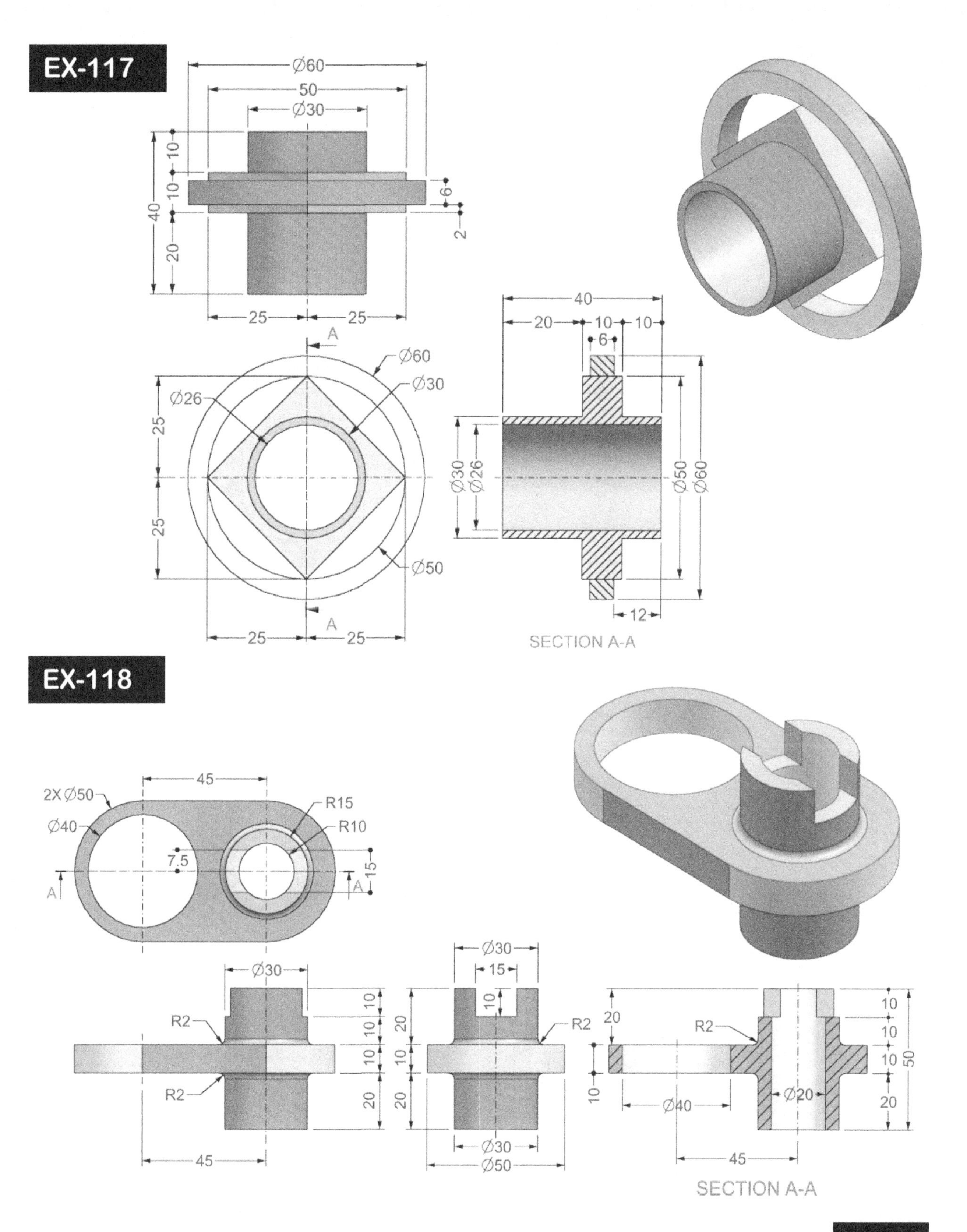
EX-117
Ø60
50
Ø30
10
10
40
20
6
2
25
25
A
Ø60
Ø30
Ø26
25
25
Ø50
A
25
25
40
20
10
10
6
Ø30
Ø26
Ø50
Ø60
12
SECTION A-A
EX-118
45
2X Ø50
Ø40
R15
R10
7.5
15
A
A
Ø30
R2
R2
10
10
10
20
45
Ø30
15
10
20
10
20
R2
Ø30
Ø50
20
R2
10
10
10
10
50
20
Ø40
Ø20
45
SECTION A-A

EX-119

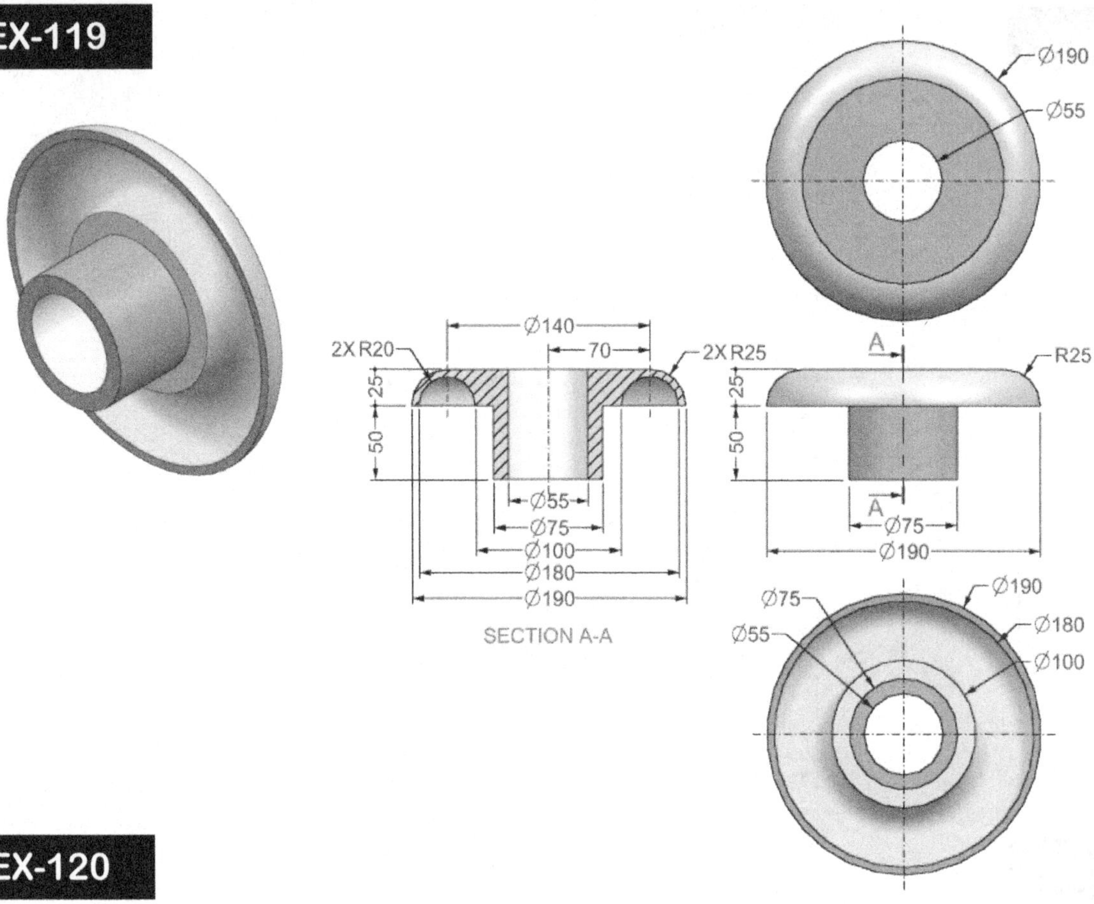

Ø190
Ø55
Ø140
70
2X R20
2X R25
25
50
Ø55
Ø75
Ø100
Ø180
Ø190

SECTION A-A

A
R25
25
50
A
Ø75
Ø190

Ø75
Ø190
Ø55
Ø180
Ø100

EX-120

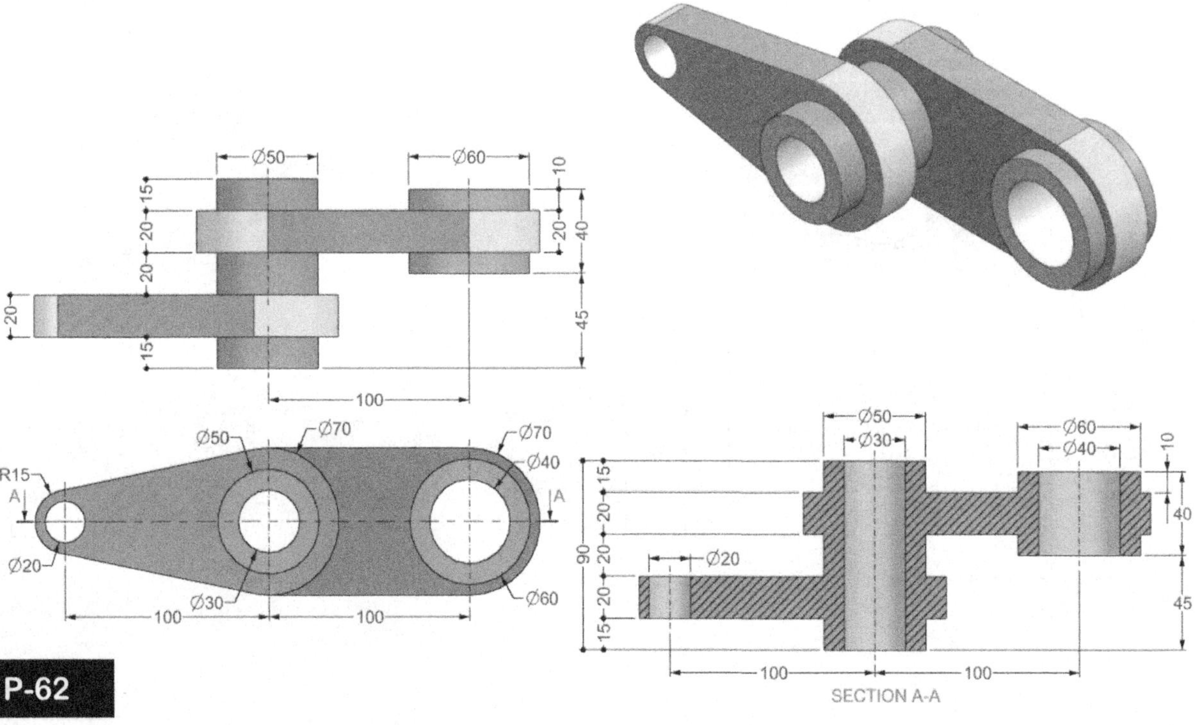

Ø50
Ø60
10
15
20
20
20
40
20
45
15
100

R15
A
Ø50
Ø70
Ø70
Ø40
A
Ø20
Ø30
Ø60
100
100

Ø50
Ø30
Ø60
Ø40
10
15
20
40
90
20
Ø20
20
45
15
100
100

SECTION A-A

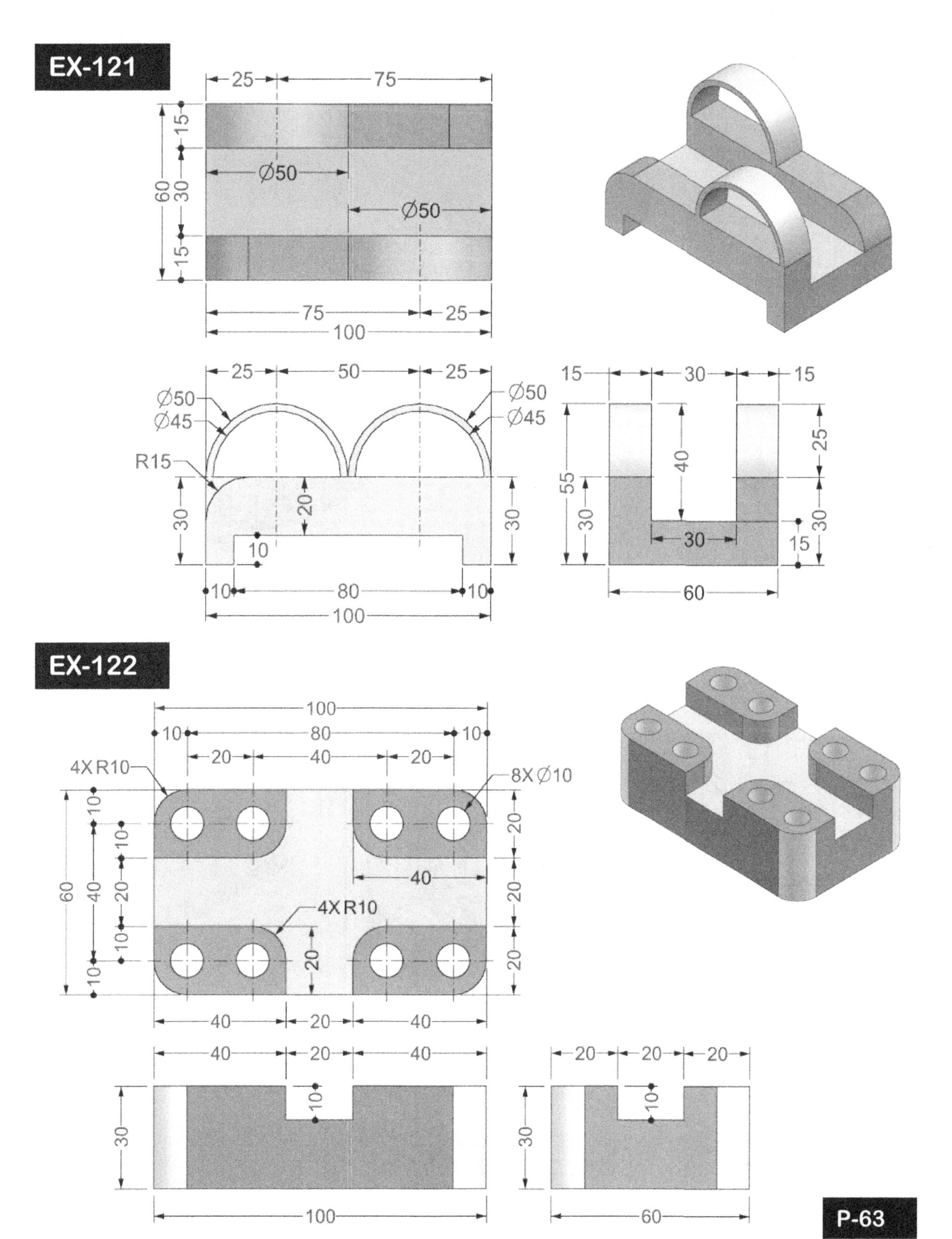
EX-121
25
75
15
30
15
60
Ø50
Ø50
75
25
100
25
50
25
Ø50
Ø45
Ø50
Ø45
R15
30
20
10
30
10
80
10
100
15
30
15
55
40
25
30
30
15
30
60
EX-122
100
10
80
10
20
40
20
4X R10
8X Ø10
10
10
20
10
10
60
40
20
40
20
20
4X R10
20
40
20
40
40
20
40
10
30
100
20
20
20
10
30
60

EX-123

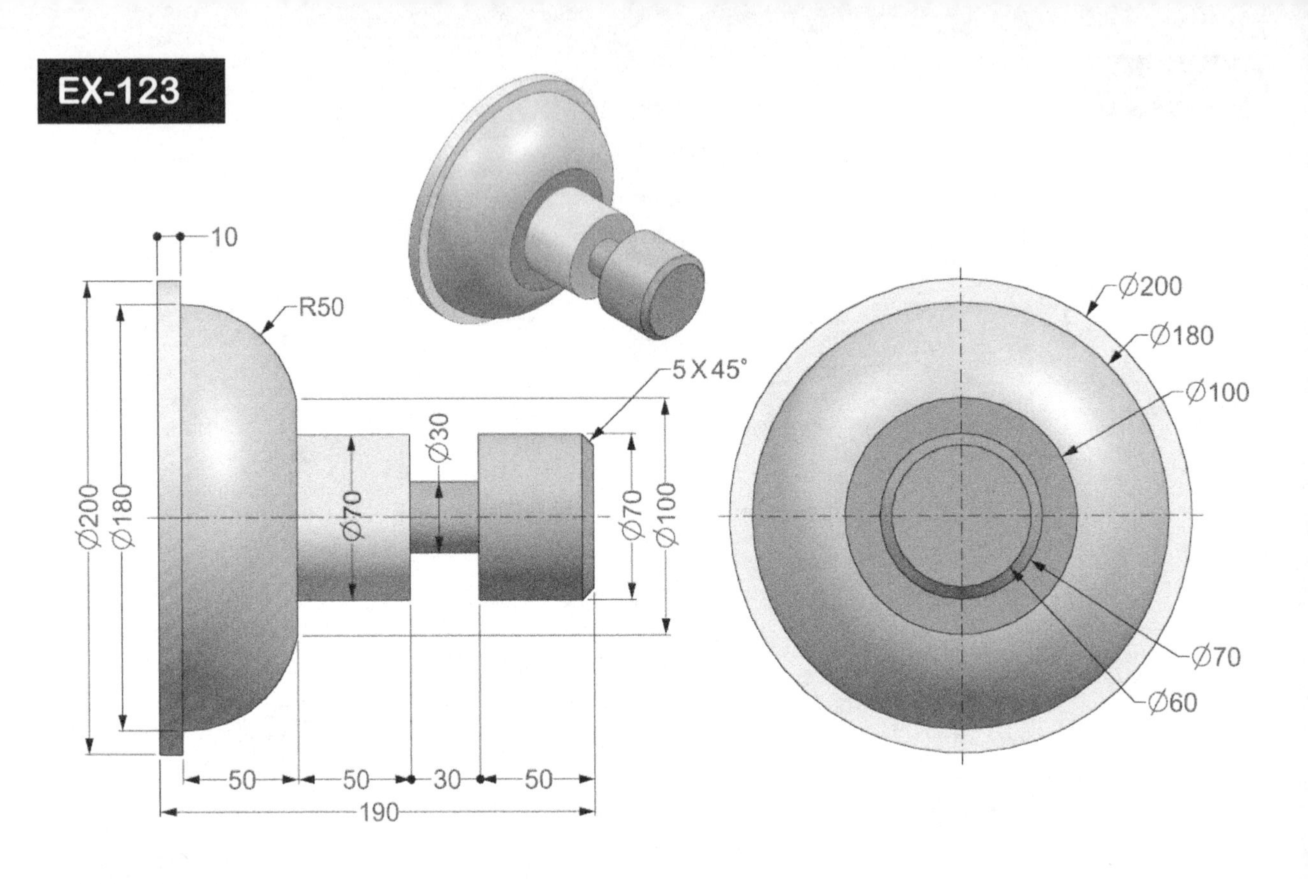
10
R50
5 X 45°
Ø30
Ø200
Ø180
Ø70
Ø70
Ø100
Ø200
Ø180
Ø100
Ø70
Ø60
50
50
30
50
190

EX-124

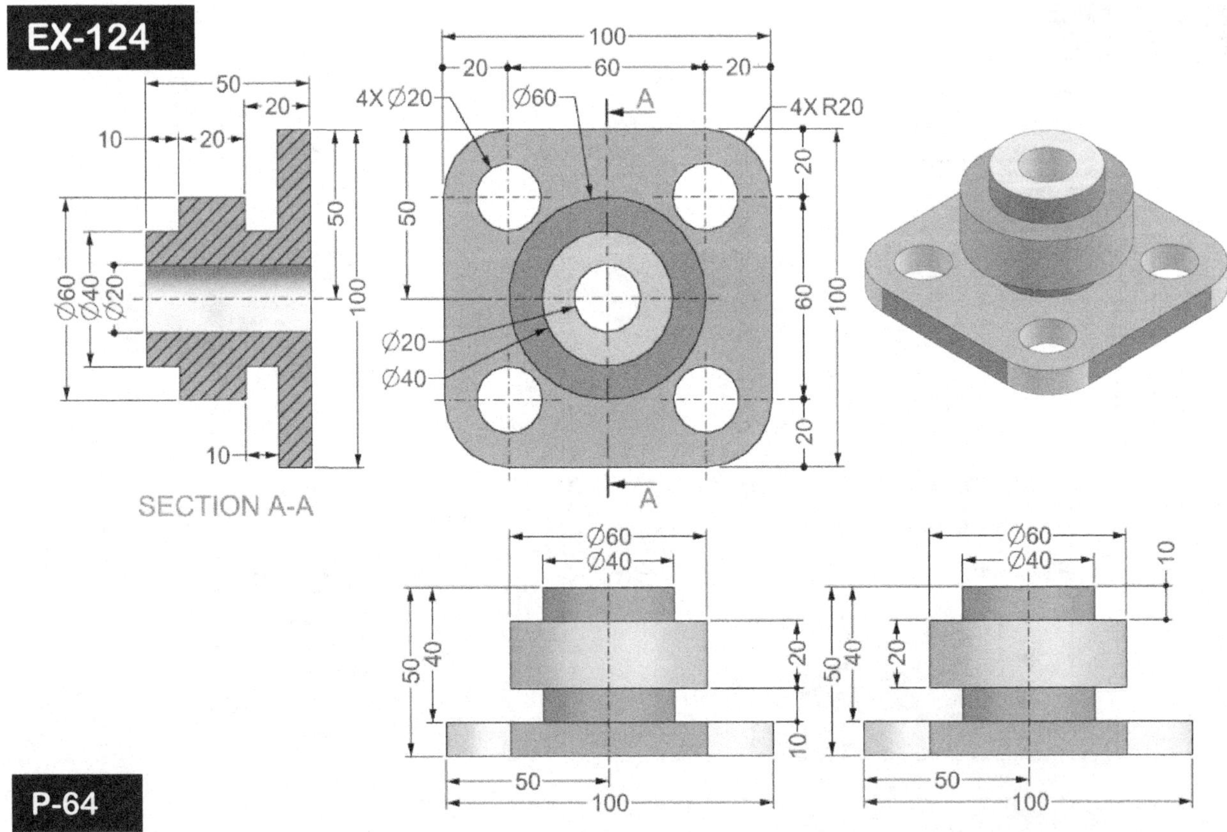
50
20
10
20
50
100
Ø60
Ø40
Ø20
10
SECTION A-A
100
20
60
20
4X Ø20
Ø60
A
4X R20
50
20
60
100
20
Ø20
Ø40
A
Ø60
Ø40
50
40
20
10
50
100
Ø60
Ø40
10
50
40
20
50
100

EX-125

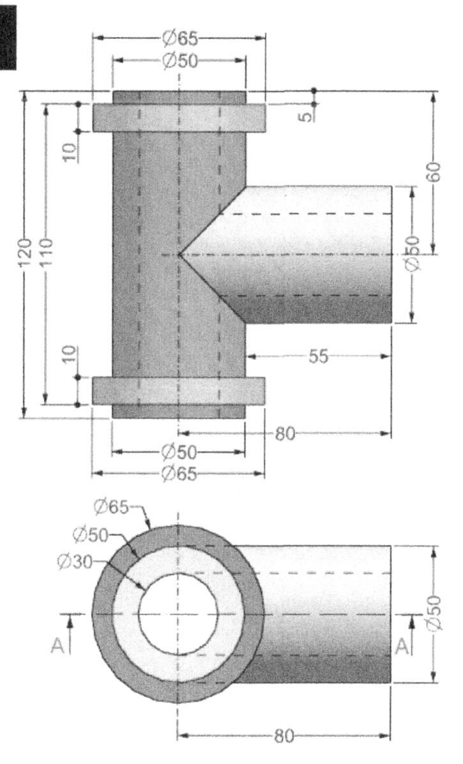

Ø65
Ø50
5
60
10
120
110
Ø50
10
55
80
Ø50
Ø65
Ø65
Ø50
Ø30
A
A
Ø50
80

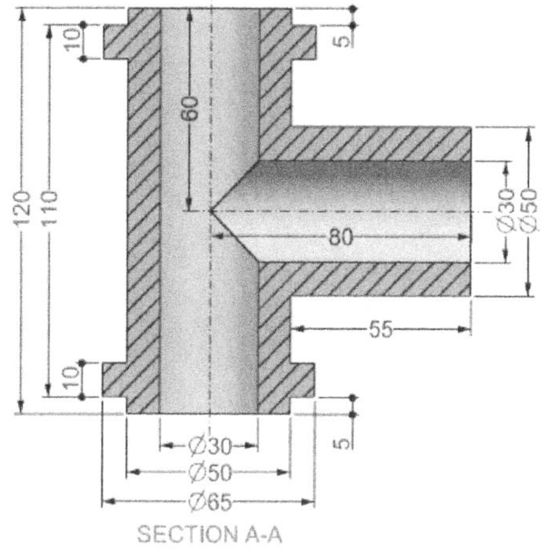

10
5
60
120
110
Ø30
Ø50
80
55
10
5
Ø30
Ø50
Ø65
SECTION A-A

EX-126

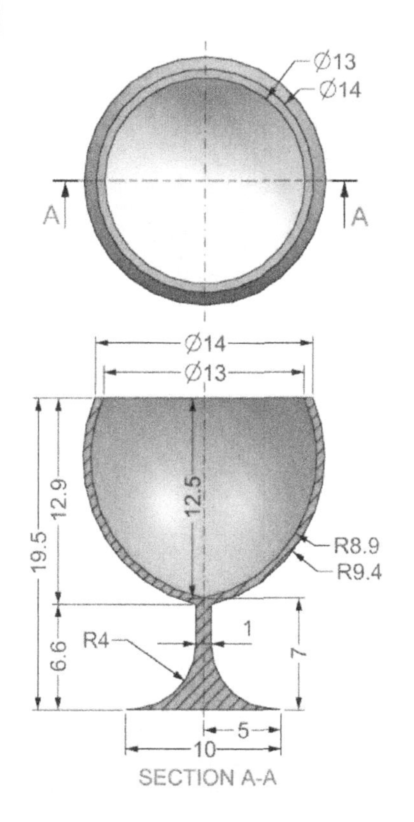

Ø13
Ø14
A
A
Ø14
Ø13
19.5
12.9
12.5
R8.9
R9.4
1
R4
6.6
7
5
10
SECTION A-A

EX-127

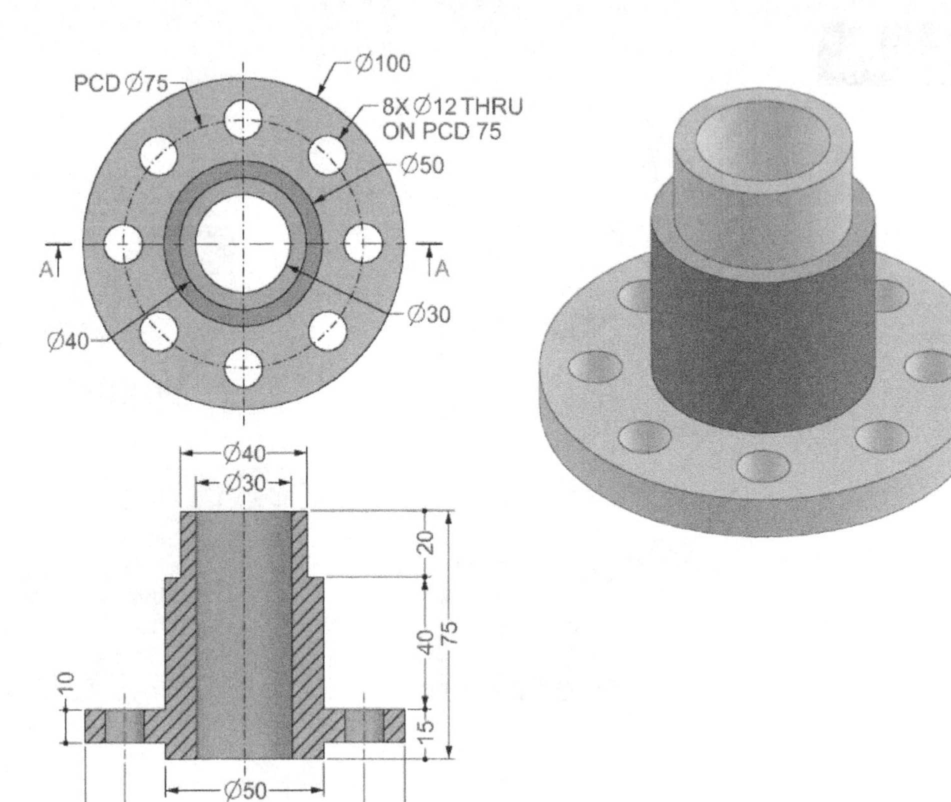

PCD Ø75
Ø100
8X Ø12 THRU
ON PCD 75
Ø50
A
A
Ø30
Ø40
Ø40
Ø30
20
40
75
10
15
Ø50
75
Ø100
SECTION A-A

EX-128

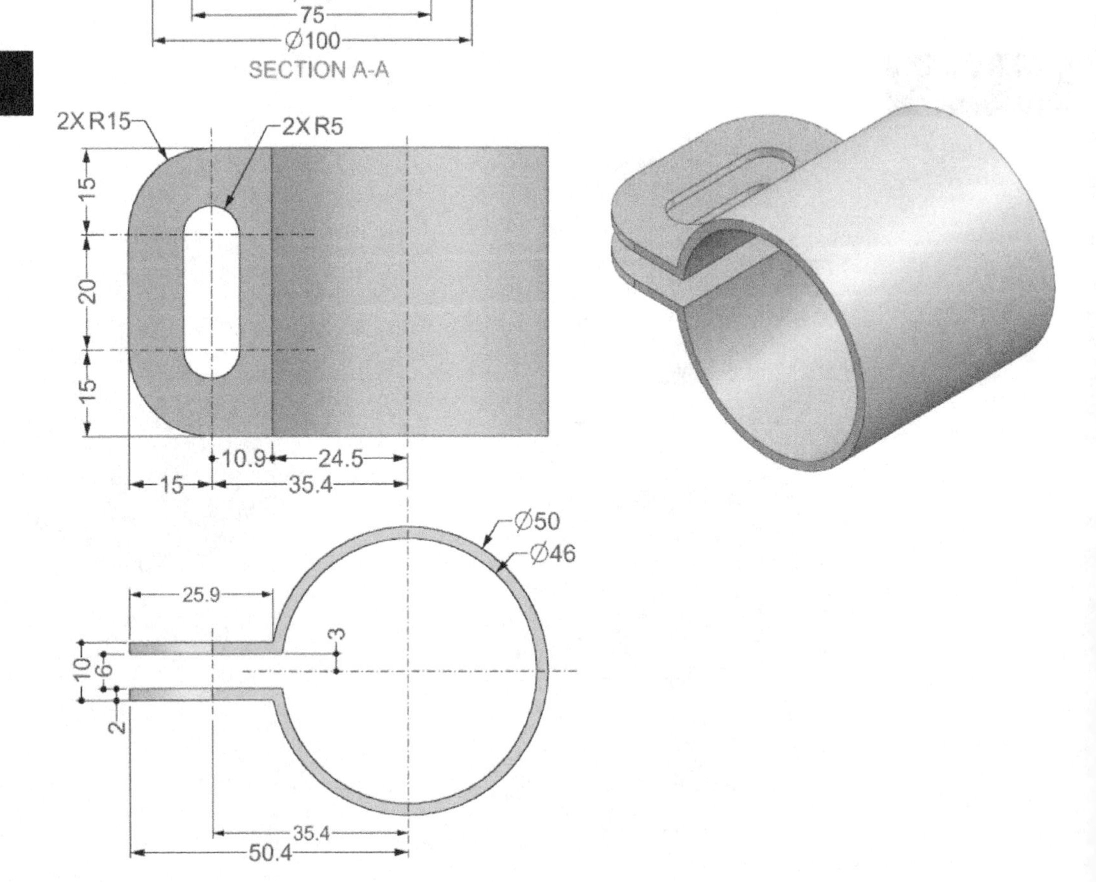

2X R15
2X R5
15
20
15
10.9
24.5
15
35.4
Ø50
Ø46
25.9
3
10
6
2
35.4
50.4

EX-129

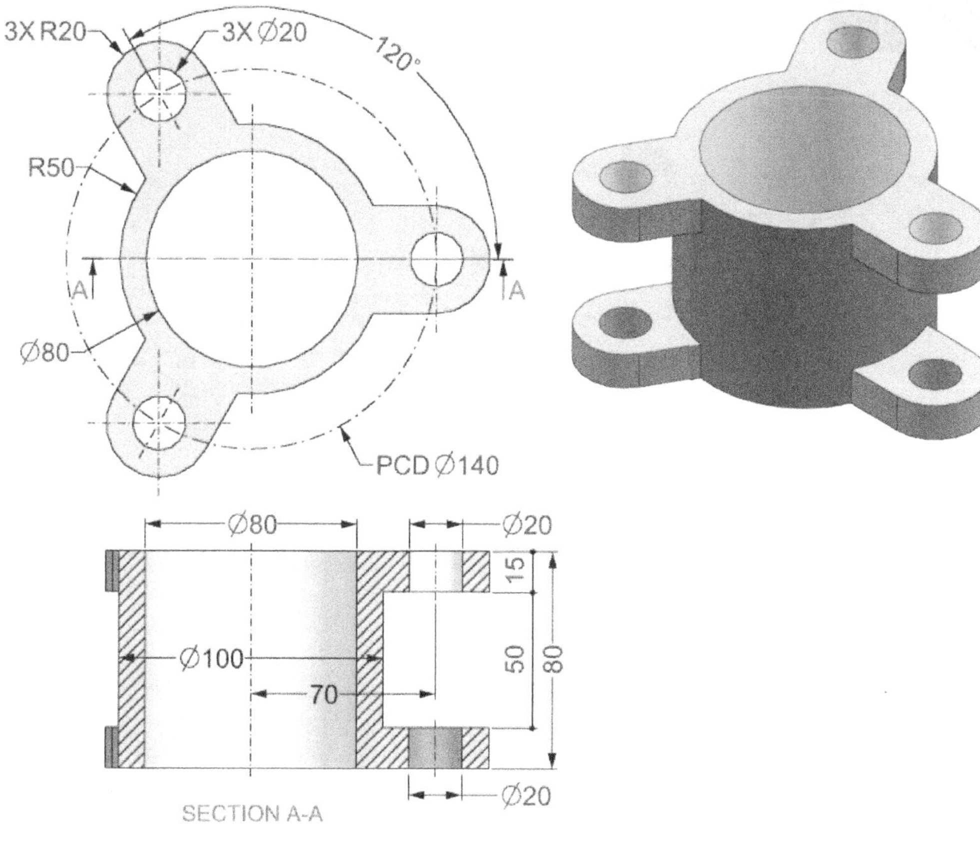
3X R20
3X Ø20
120°
R50
A
A
Ø80
PCD Ø140
Ø80
Ø20
15
Ø100
50
80
70
Ø20
SECTION A-A

EX-130

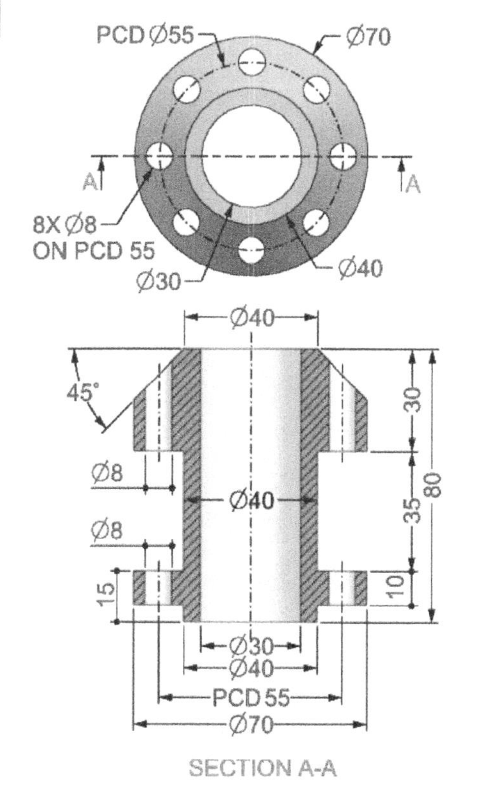
PCD Ø55
Ø70
A
A
8X Ø8
ON PCD 55
Ø30
Ø40
Ø40
45°
30
Ø8
Ø40
35
80
Ø8
15
10
Ø30
Ø40
PCD 55
Ø70
SECTION A-A

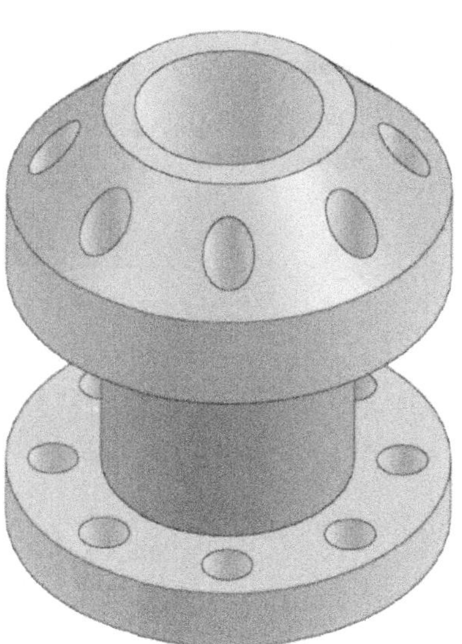

EX-131

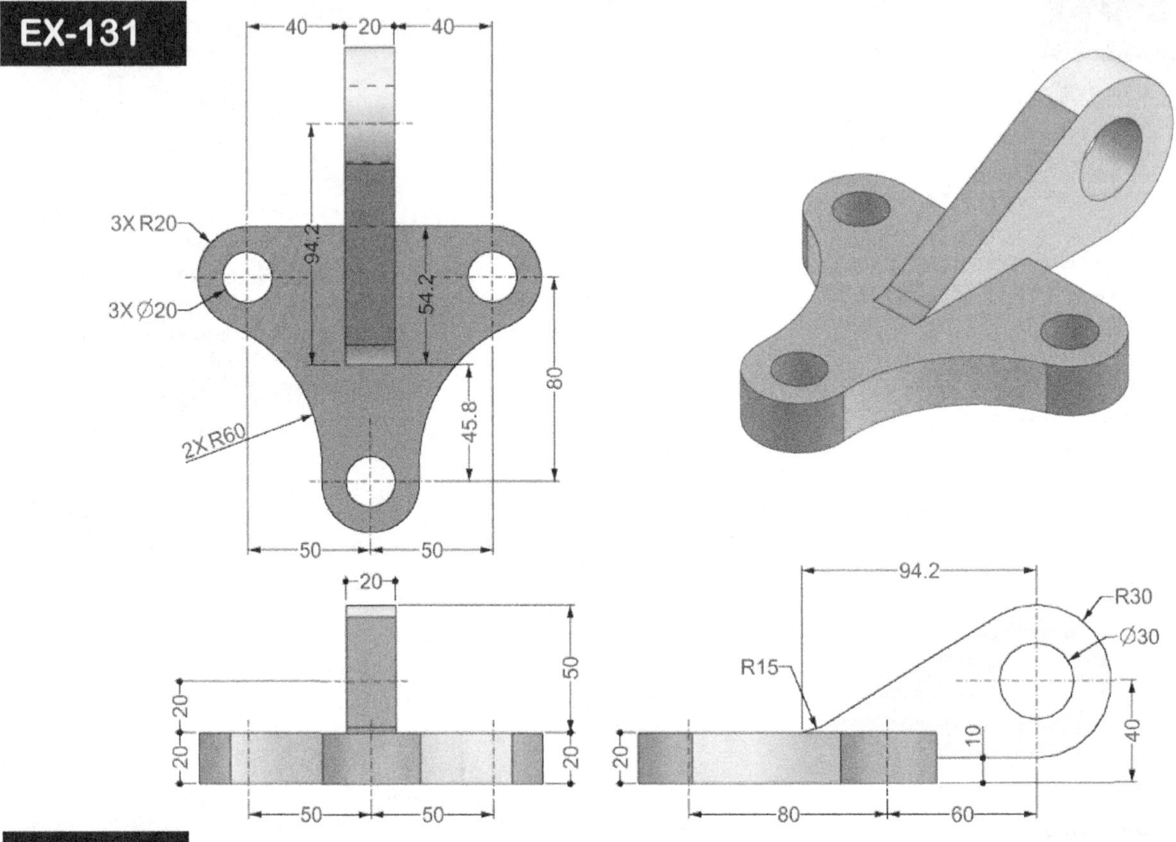

40
20
40
3X R20
3X Ø20
94.2
54.2
80
45.8
2X R60
50
50
20
50
20
20
20
50
50
94.2
R30
Ø30
R15
20
10
40
80
60

EX-132

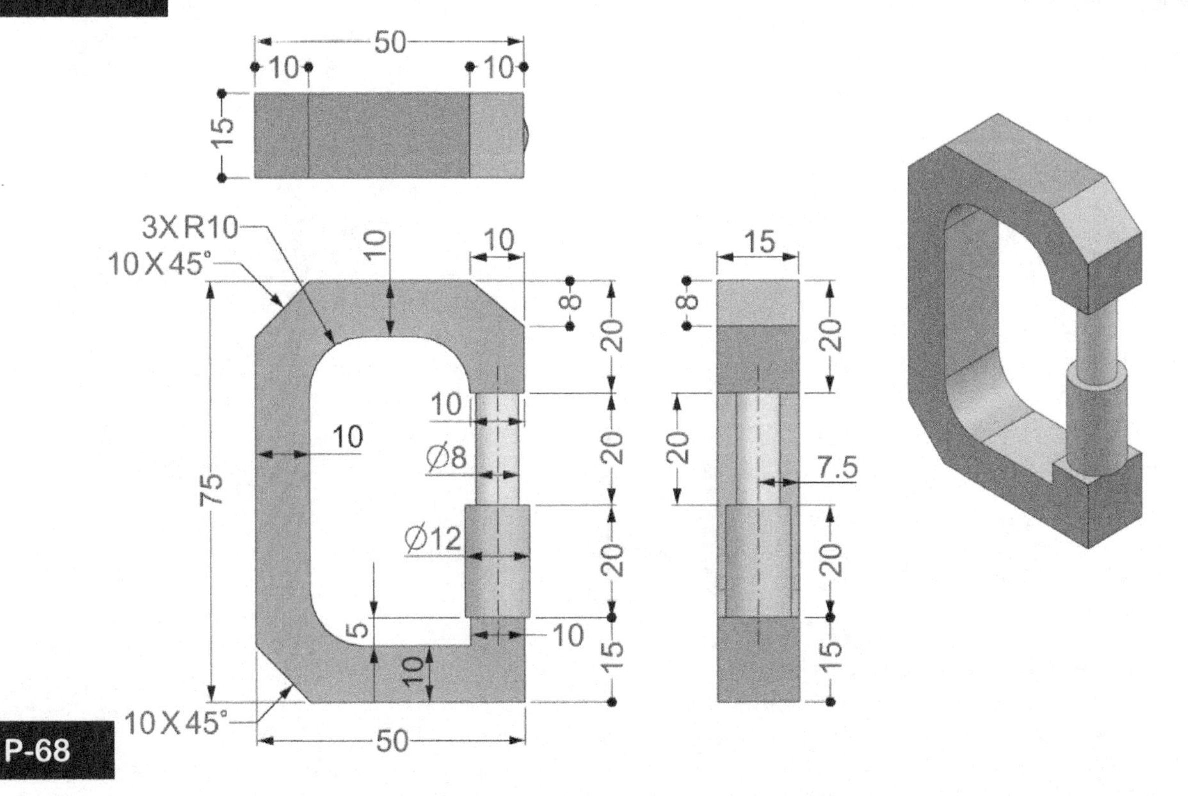

50
10
10
15
3X R10
10 X 45°
10
10
8
20
15
8
20
10
10
Ø8
20
20
7.5
75
Ø12
20
20
5
10
15
15
10
10 X 45°
50

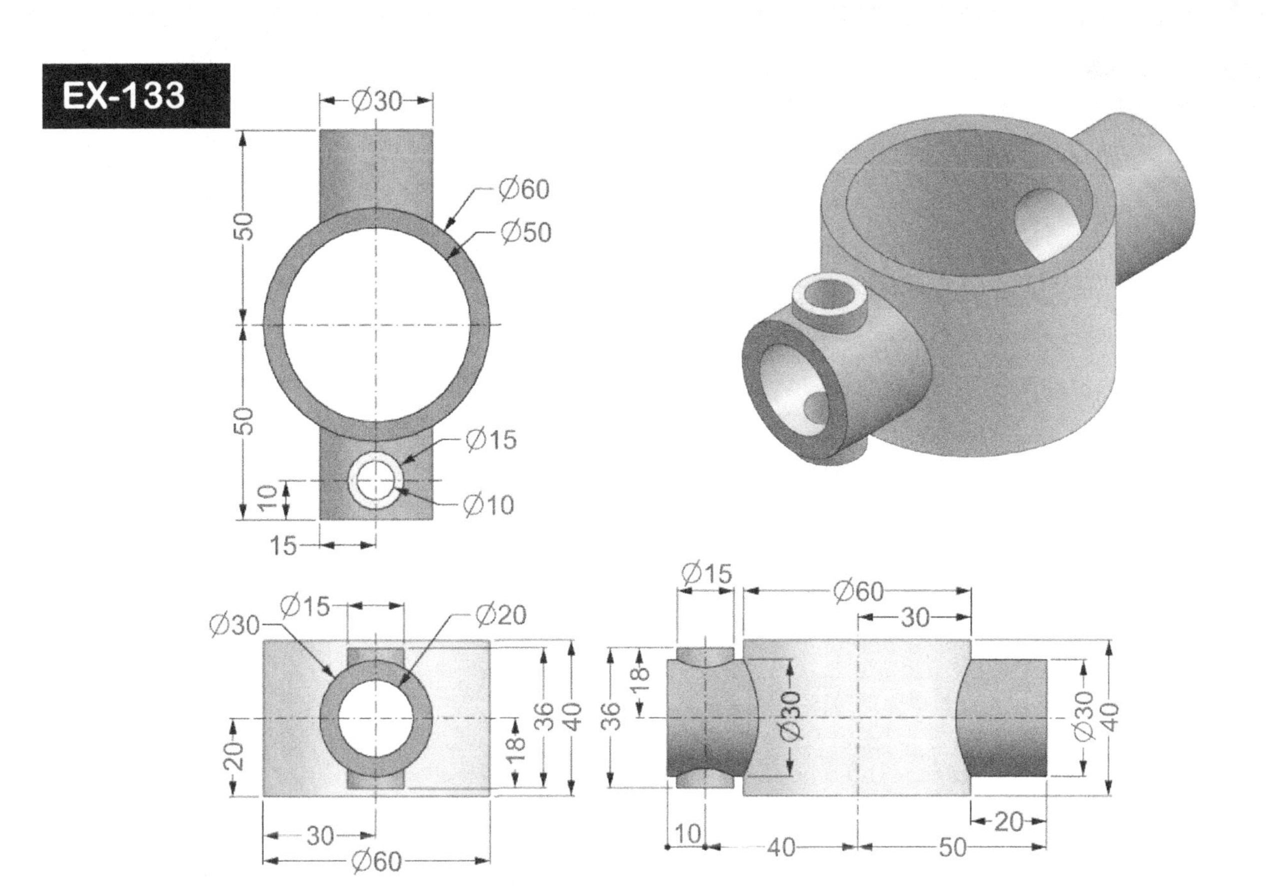

EX-134

80
70
50
2 X 45°
R30
Ø40
Ø120
Ø60
A
A
Ø120
Ø60
Ø93.2
Ø60
Ø120
5
5
10
10
Ø120
Ø60
Ø40
5
2 X 45°
10
80
Ø93.2
50
70
Ø120

SECTION A-A

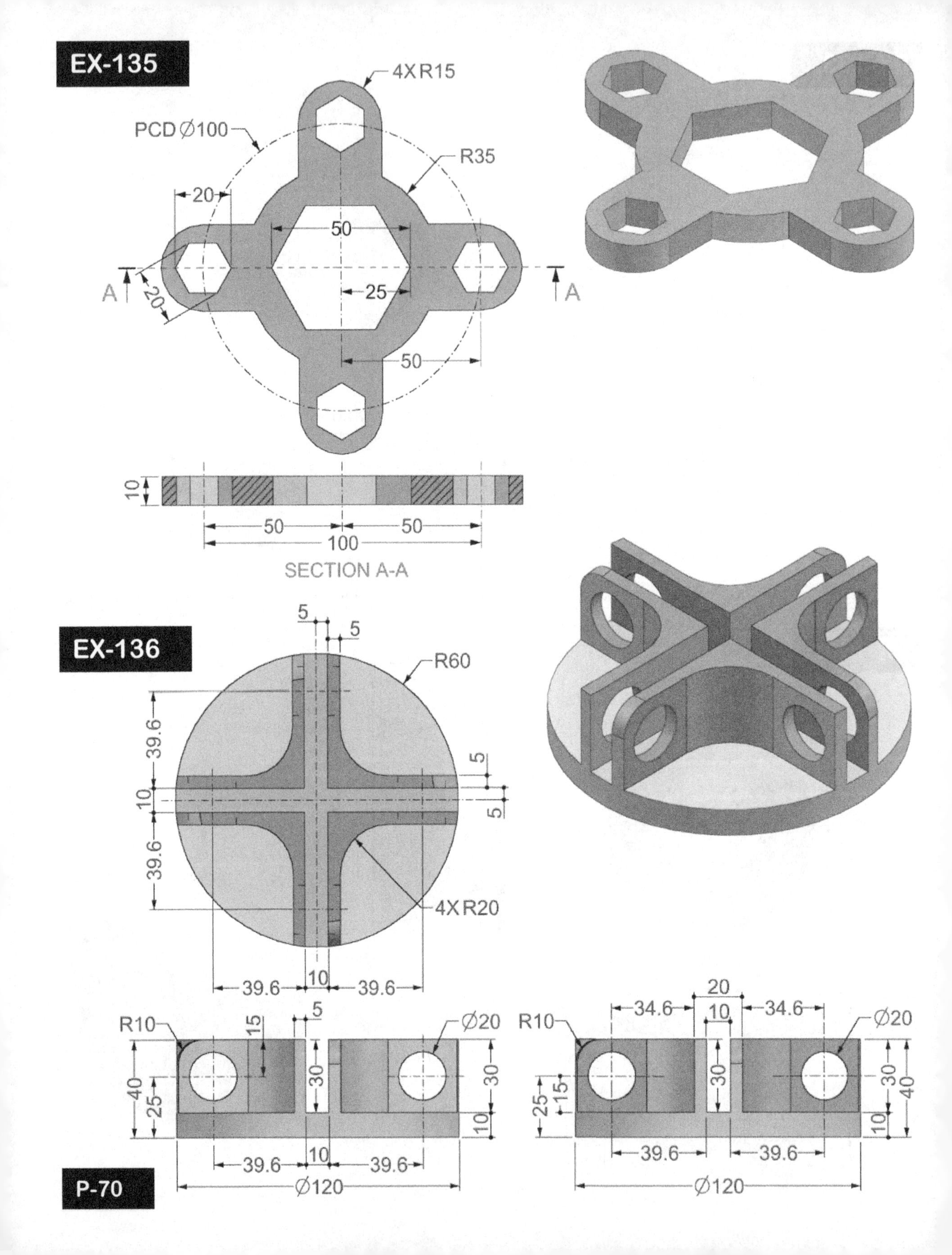
EX-135
4X R15
PCD Ø100
R35
20
50
A
A
20
25
50
10
50
50
100
SECTION A-A
EX-136
5
5
R60
39.6
5
10
5
39.6
4X R20
39.6
10
39.6
R10
15
5
Ø20
40
25
30
30
10
39.6
10
39.6
Ø120
20
34.6
10
34.6
R10
Ø20
25
15
30
30
40
10
39.6
39.6
Ø120
P-70

EX-137

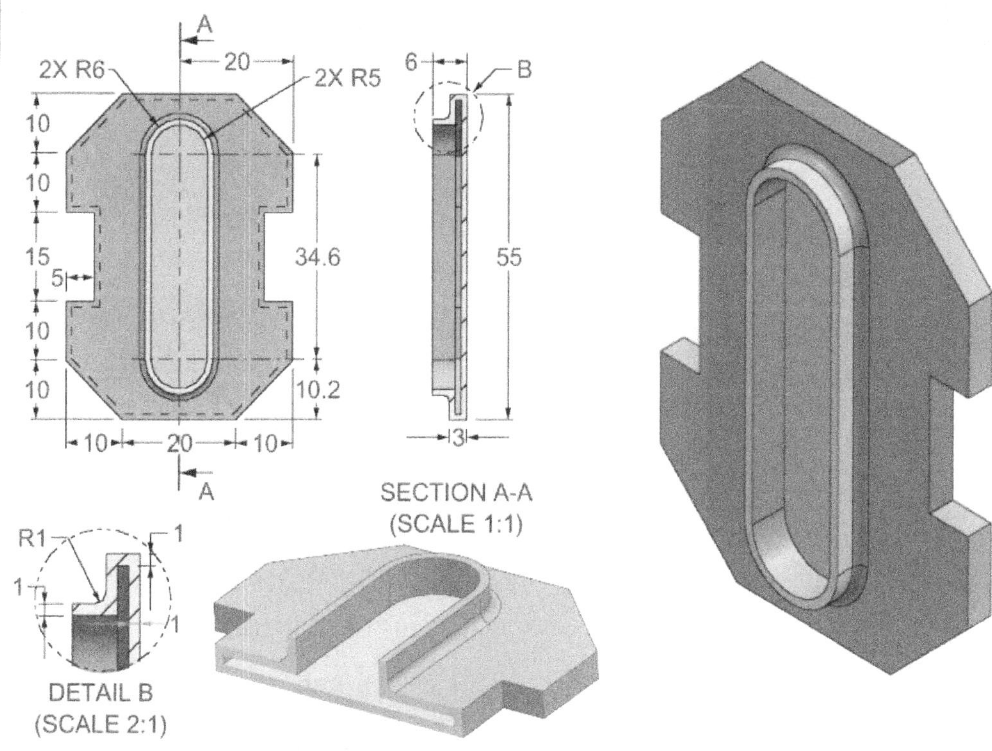

SHELL THICKNESS = 1MM
ALL INSIDE WALL THICKNESS

EX-138

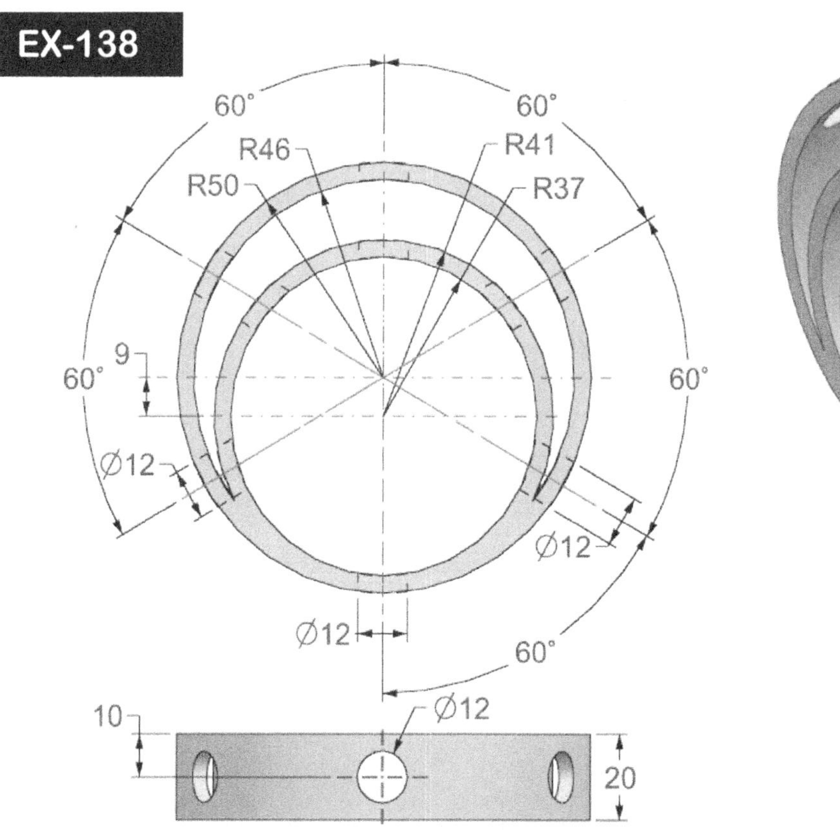

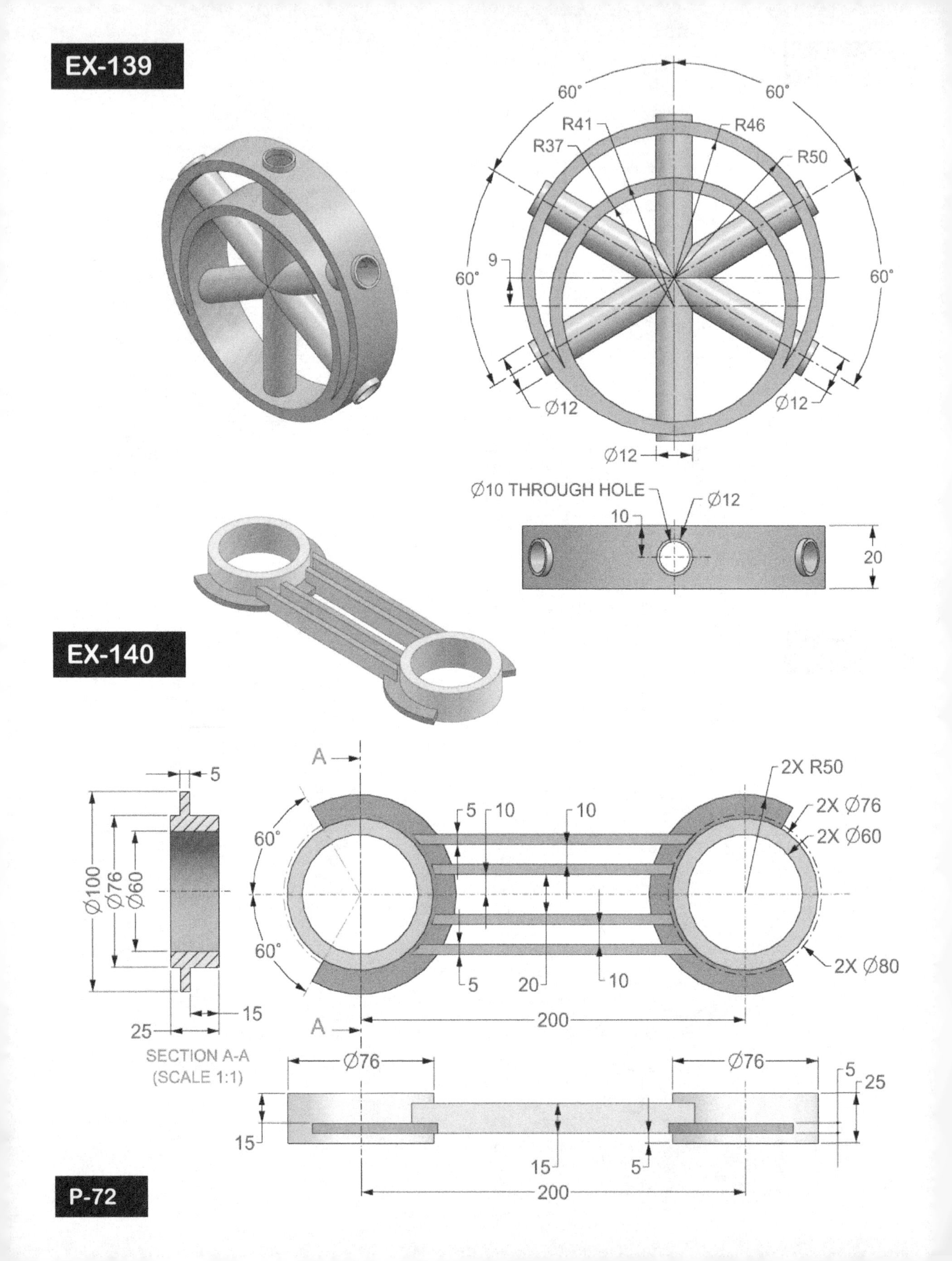

EX-139
60°
60°
R41
R46
R37
R50
9
60°
60°
Ø12
Ø12
Ø12
Ø10 THROUGH HOLE
Ø12
10
20
EX-140
A
A
5
2X R50
5
10
10
2X Ø76
2X Ø60
60°
Ø100
Ø76
Ø60
60°
5
20
10
2X Ø80
15
25
200
SECTION A-A
(SCALE 1:1)
Ø76
Ø76
5
25
15
15
5
200
P-72

EX-141

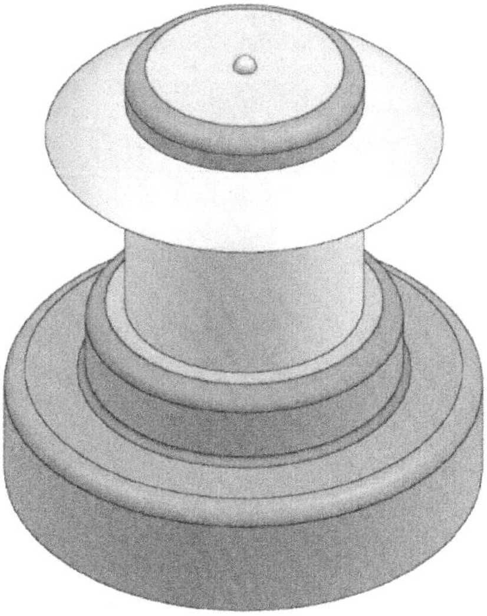

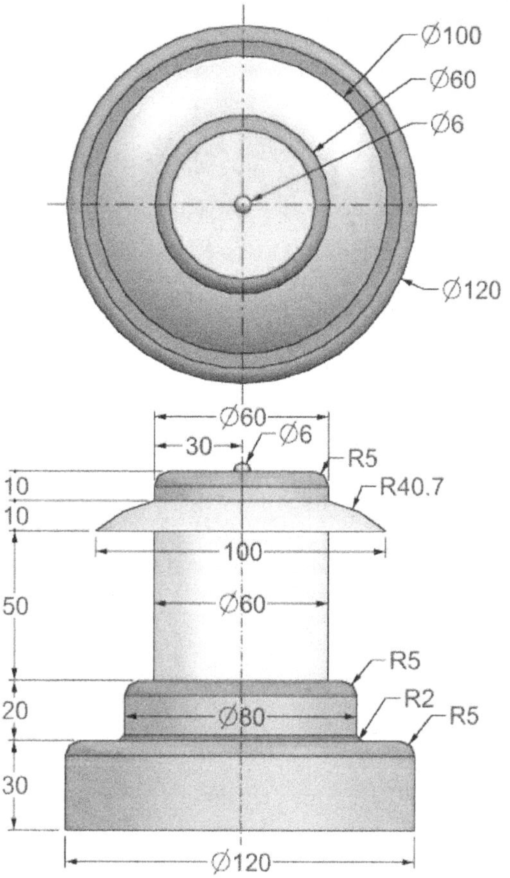

Ø100
Ø60
Ø6
Ø120

Ø60
30
Ø6
R5
R40.7
10
10
100
50
Ø60
R5
20
Ø80
R2
R5
30
Ø120

EX-142

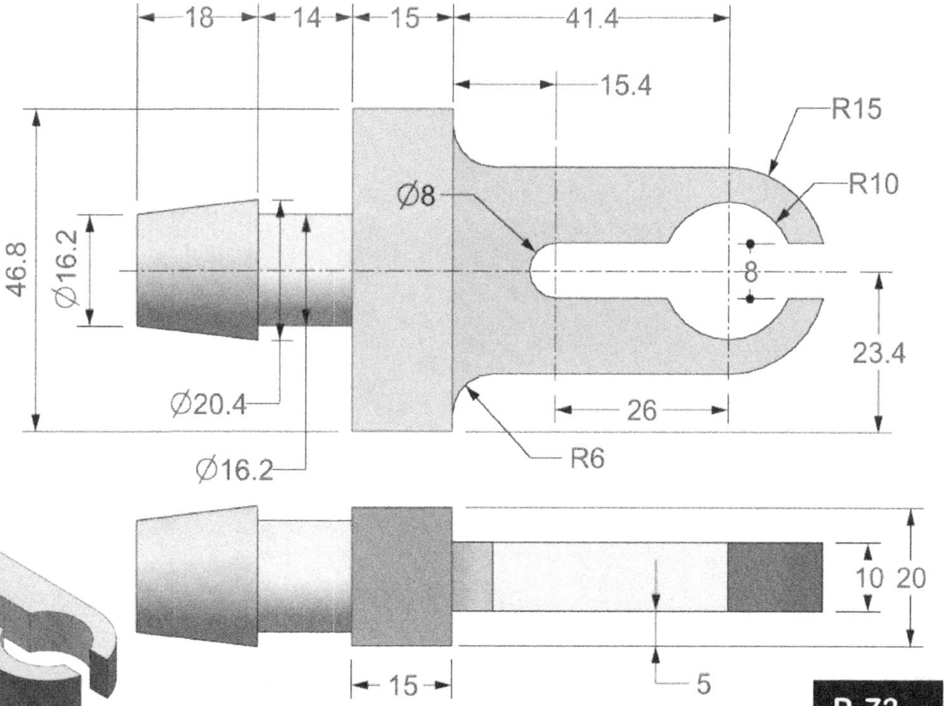

18
14
15
41.4
15.4
R15
R10
Ø8
46.8
Ø16.2
8
23.4
Ø20.4
26
Ø16.2
R6
10
20
15
5

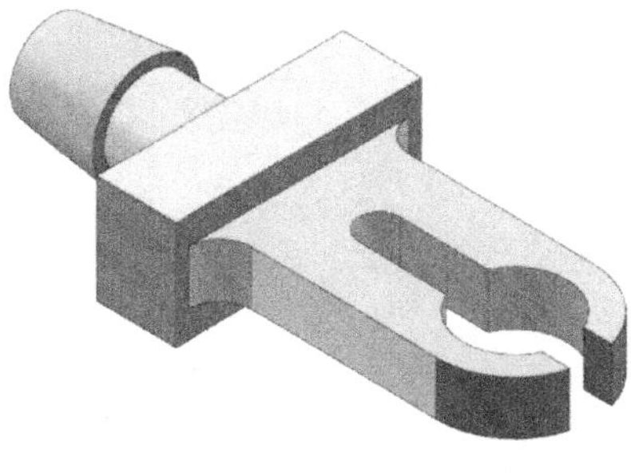

EX-143

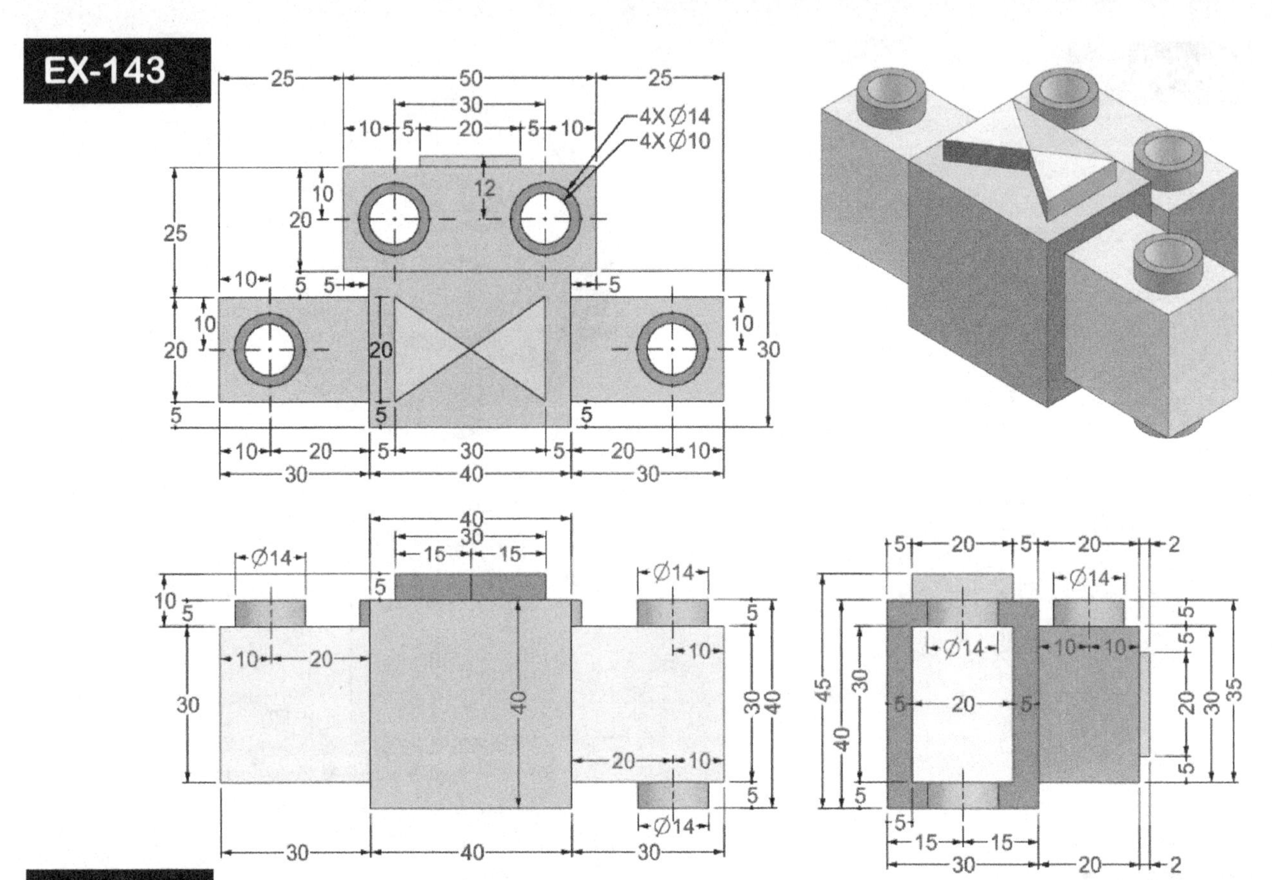
4X Ø14
4X Ø10

EX-144

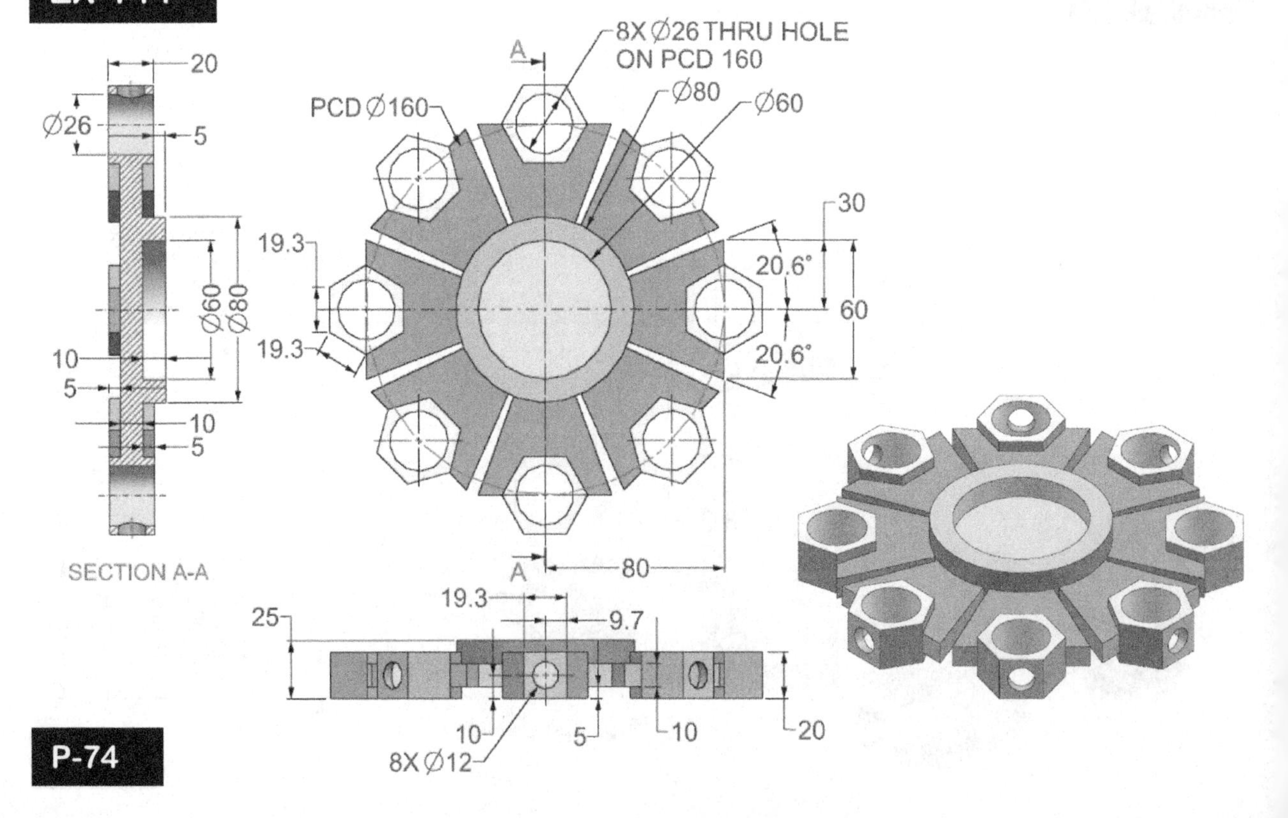
8X Ø26 THRU HOLE
ON PCD 160
PCD Ø160
Ø80
Ø60
SECTION A-A
8X Ø12

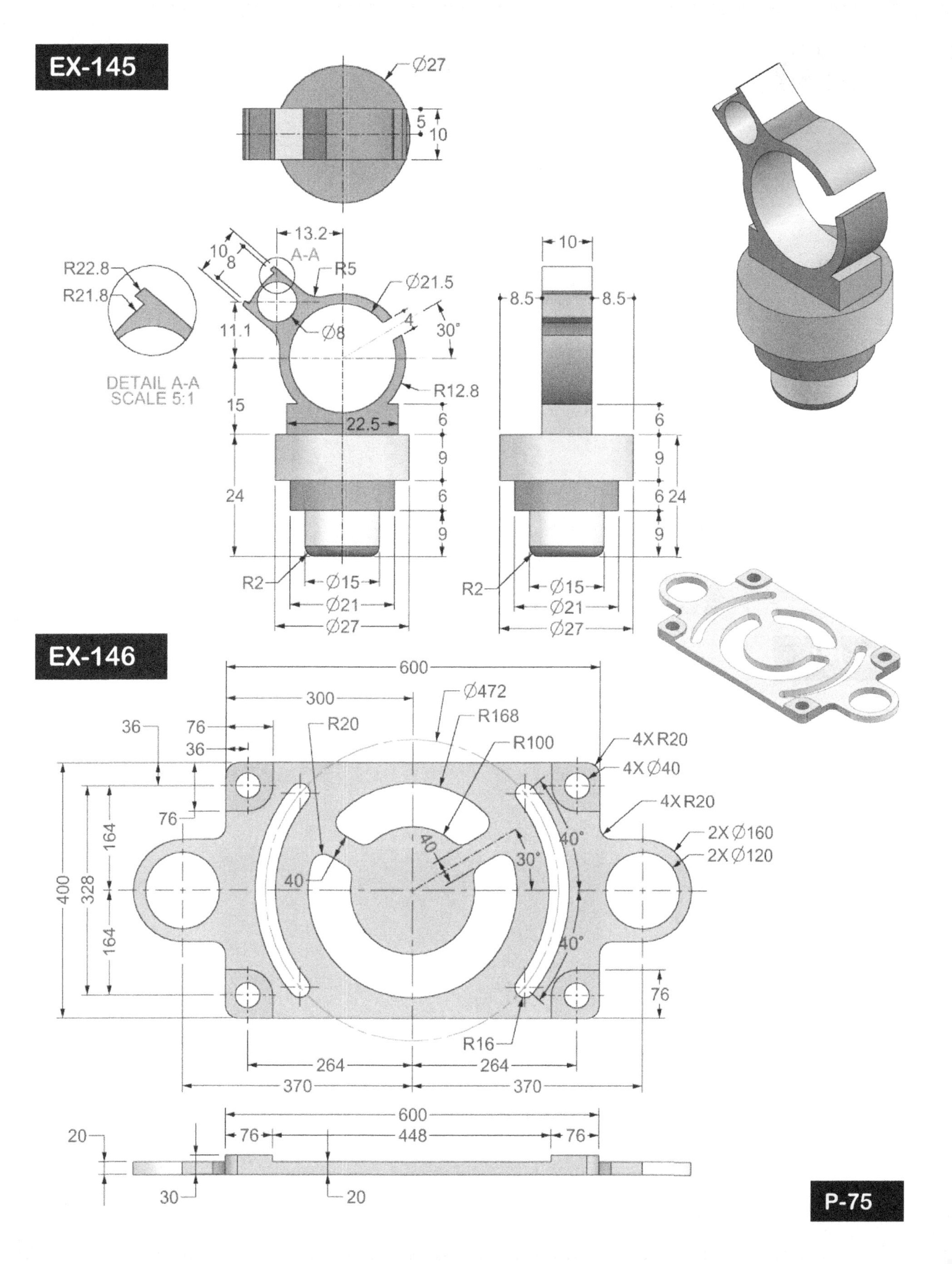
EX-145
Ø27
5
10
13.2
10
8
A-A
R5
Ø21.5
R22.8
R21.8
11.1
Ø8
4
30°
DETAIL A-A
SCALE 5:1
R12.8
15
22.5
6
9
24
6
9
R2
Ø15
Ø21
Ø27
10
8.5
8.5
EX-146
600
300
Ø472
R168
R100
R20
36
76
36
4X R20
4X Ø40
4X R20
76
164
40
40°
2X Ø160
2X Ø120
400
328
30°
40
164
40°
76
R16
264
264
370
370
600
76
448
76
20
30
20

EX-147

Ø40
Ø20
120°
120°
10
60
R10
Ø40
200
79.6
Ø20
15
R15
60

EX-148

2X Ø100
2X Ø80
Ø50
R45
R40
51.6
Ø30
A
A
100
100

Ø90
Ø50
10
40
15
10
100
100

Ø90
Ø80
Ø80
Ø50
Ø80
Ø30
15
10
10
40
15
100
100

SECTION A-A

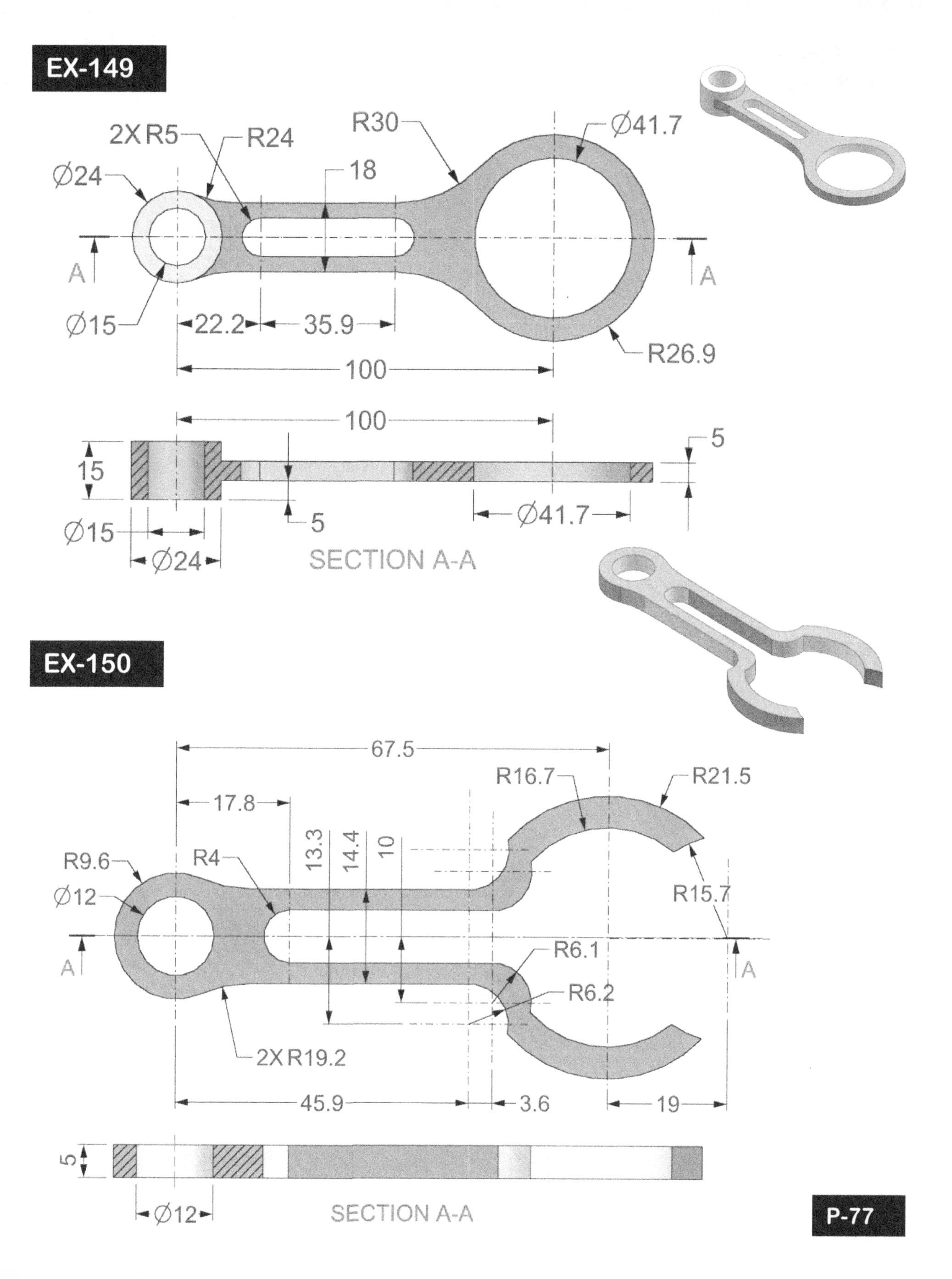
EX-149
2X R5
R24
R30
Ø41.7
Ø24
18
A
A
Ø15
22.2
35.9
R26.9
100
100
5
15
5
Ø41.7
Ø15
Ø24
SECTION A-A
EX-150
67.5
R16.7
R21.5
17.8
13.3
14.4
10
R9.6
R4
Ø12
R15.7
A
A
R6.1
R6.2
2X R19.2
45.9
3.6
19
5
Ø12
SECTION A-A

EX-151

Ø8
6.5
R1.5
10
28
R1.5
11
B-B
27
Ø10
SECTION A-A
Ø20
A
R3
35
15°
20
5
A
Ø13.3
Ø16
R8
R10
R6.7
R4
R5
1
45°
DETAIL B-B
SCALE 5:1

EX-152

Ø20
Ø36
Ø58
Ø52
Ø16
Ø36
Ø20
R8
8
20
2
135°
13.5
Ø16
R11.2
76
15.8
21.6
10.7
13
10
R6
R3
Ø16
SECTION A-A
A
58
3
R3
R2
Ø52
76
R6
A
40

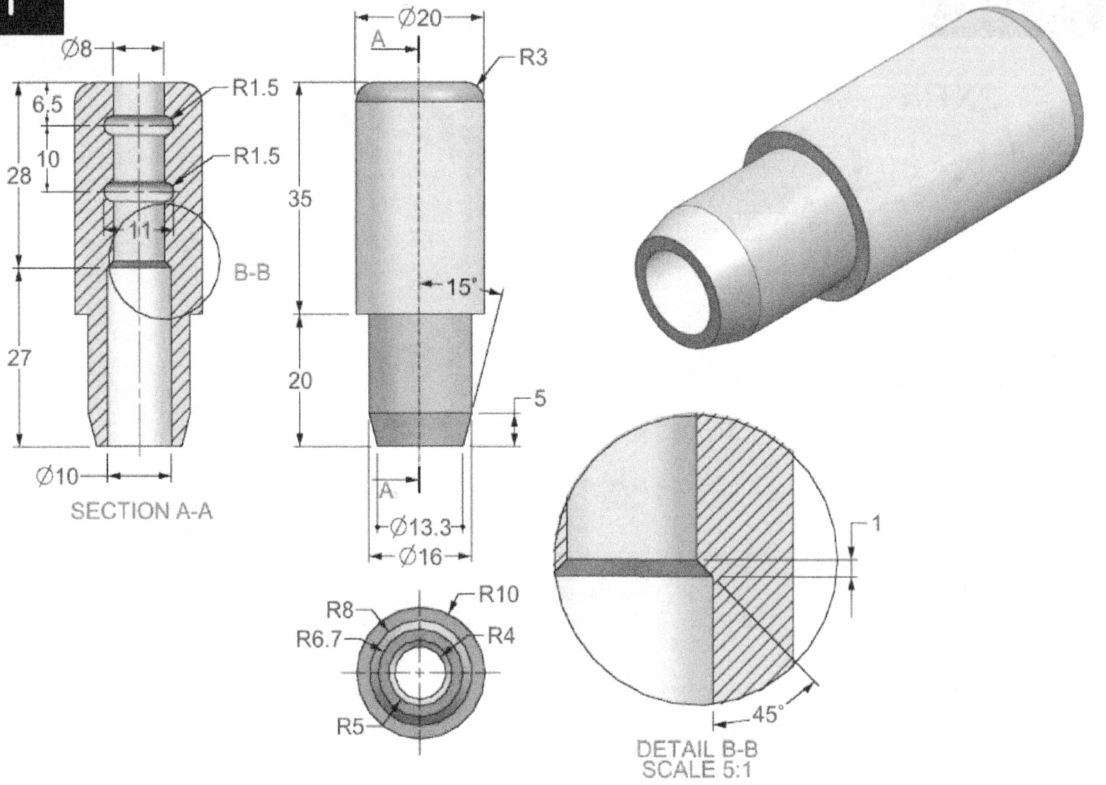

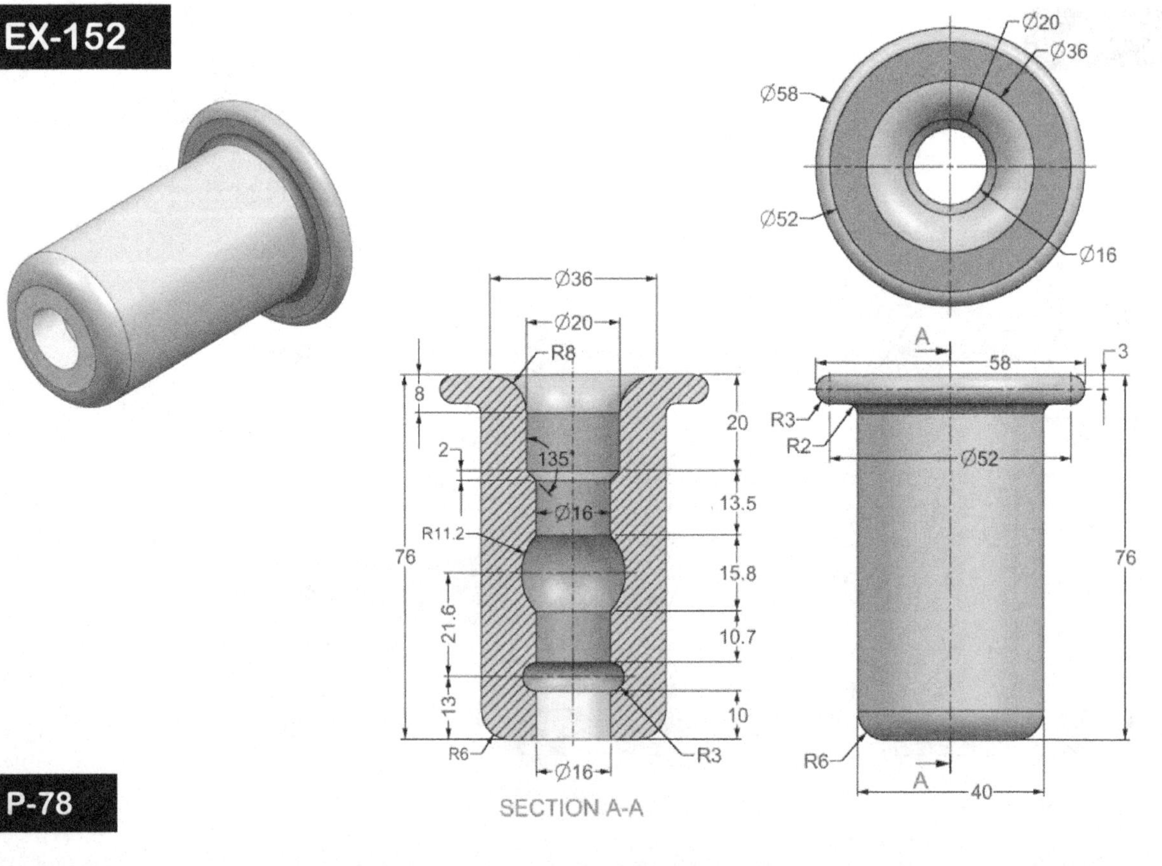

EX-153

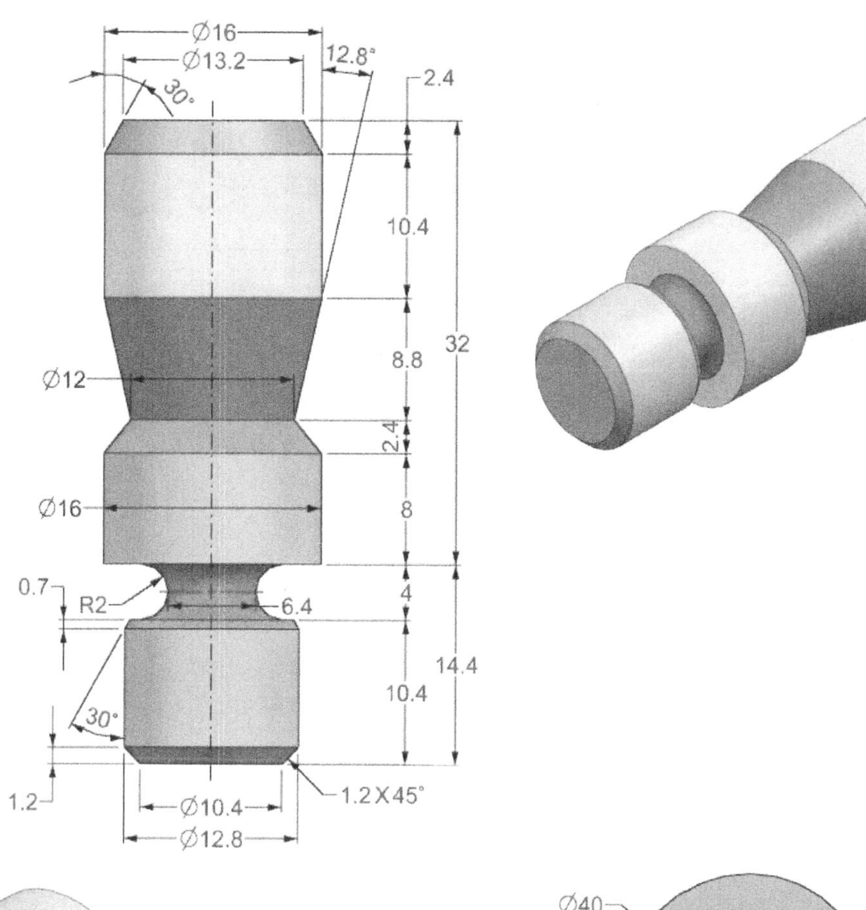

∅16
∅13.2
12.8°
2.4
30°
10.4
32
8.8
∅12
2.4
∅16
8
0.7
R2
6.4
4
14.4
10.4
30°
1.2
1.2 X 45°
∅10.4
∅12.8

EX-154

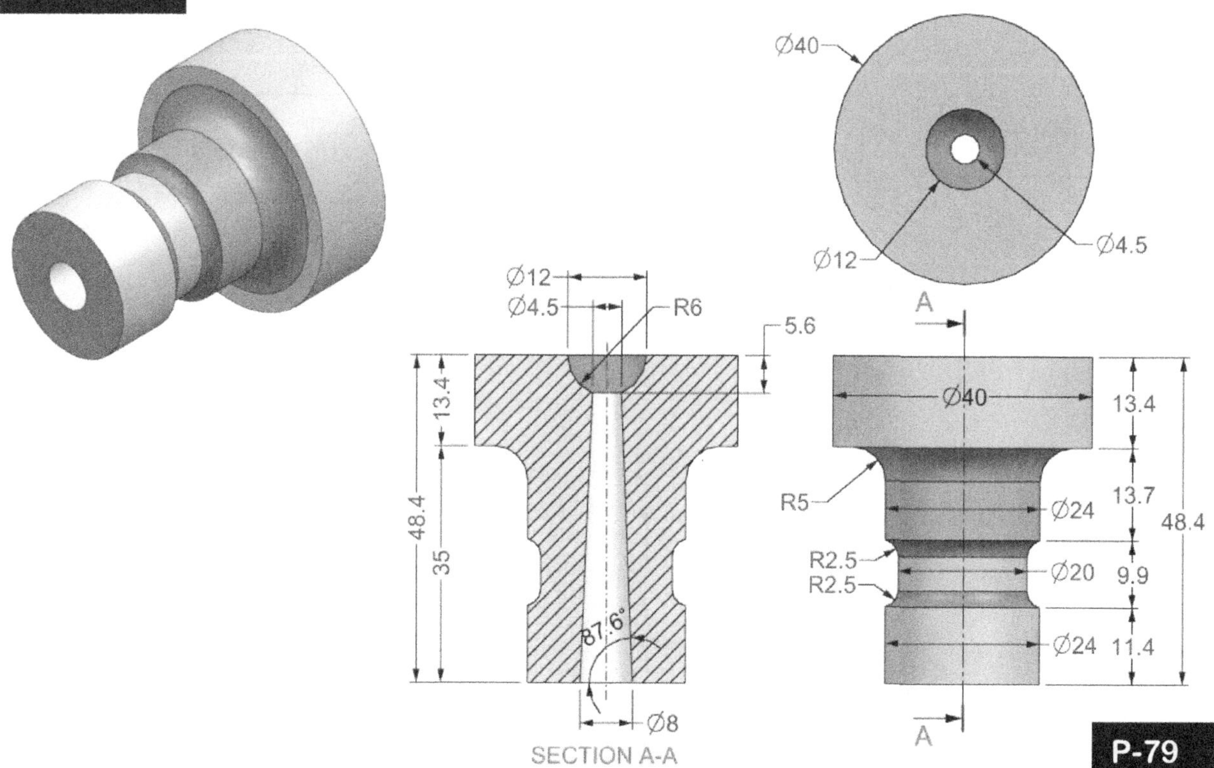

∅40
∅12
∅4.5
∅12
∅4.5
R6
5.6
13.4
48.4
35
87.6°
∅8
SECTION A-A
A
∅40
13.4
R5
13.7
∅24
48.4
R2.5
R2.5
∅20
9.9
∅24
11.4
A

EX-155

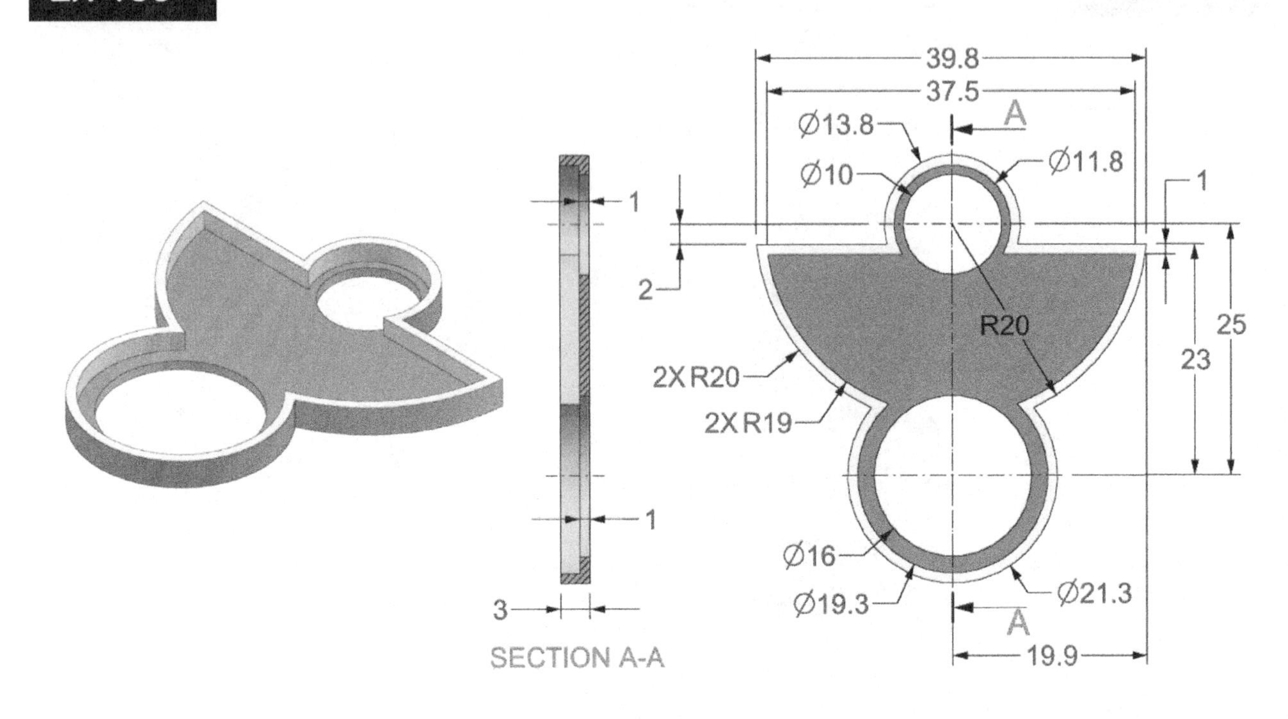

39.8
37.5
A
Ø13.8
Ø11.8
Ø10
1
2
R20
25
23
2X R20
2X R19
1
1
Ø16
Ø19.3
Ø21.3
3
SECTION A-A
19.9

EX-156

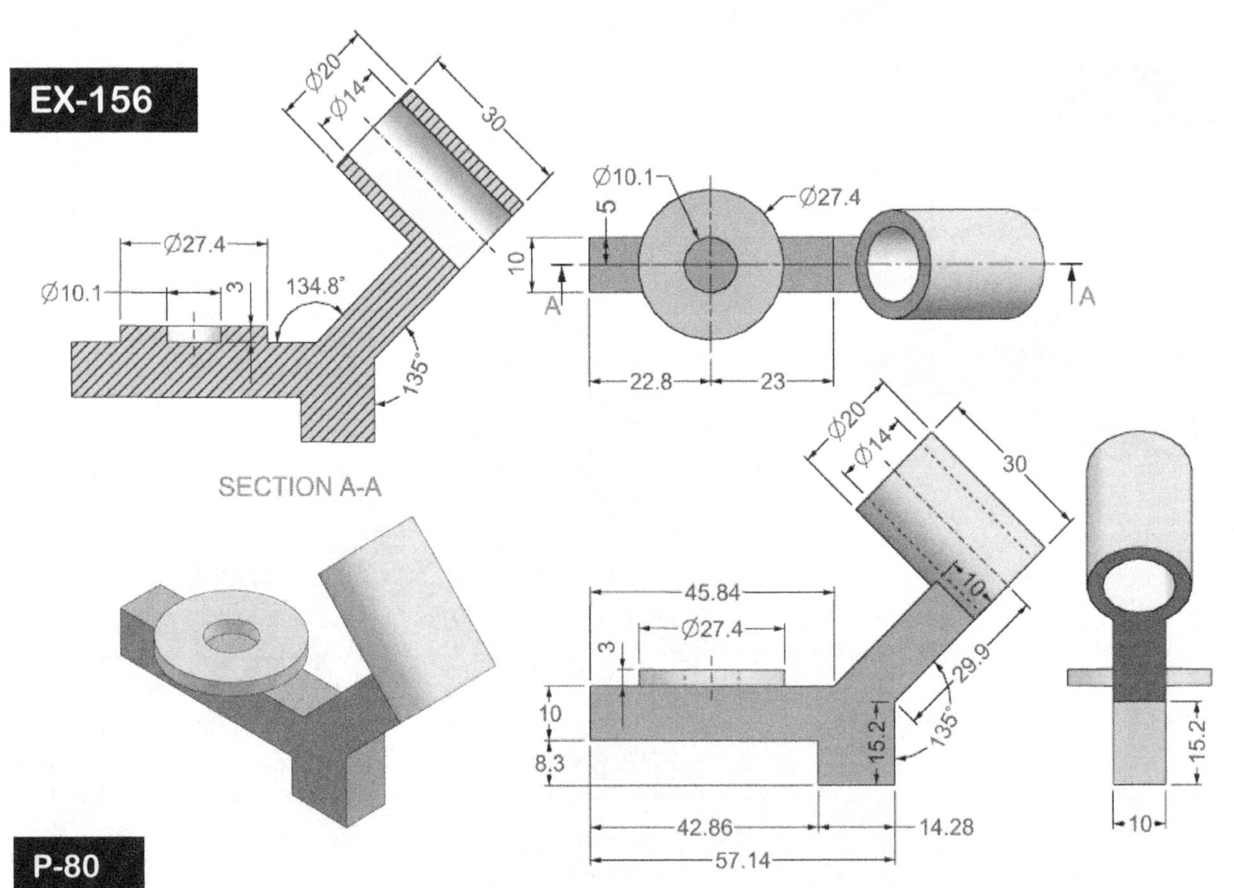

Ø20
Ø14
30
Ø27.4
Ø10.1
3
134.8°
135°
SECTION A-A
Ø10.1
Ø27.4
5
10
A
22.8
23
45.84
Ø27.4
3
10
8.3
15.2
29.9
135°
42.86
14.28
57.14
10

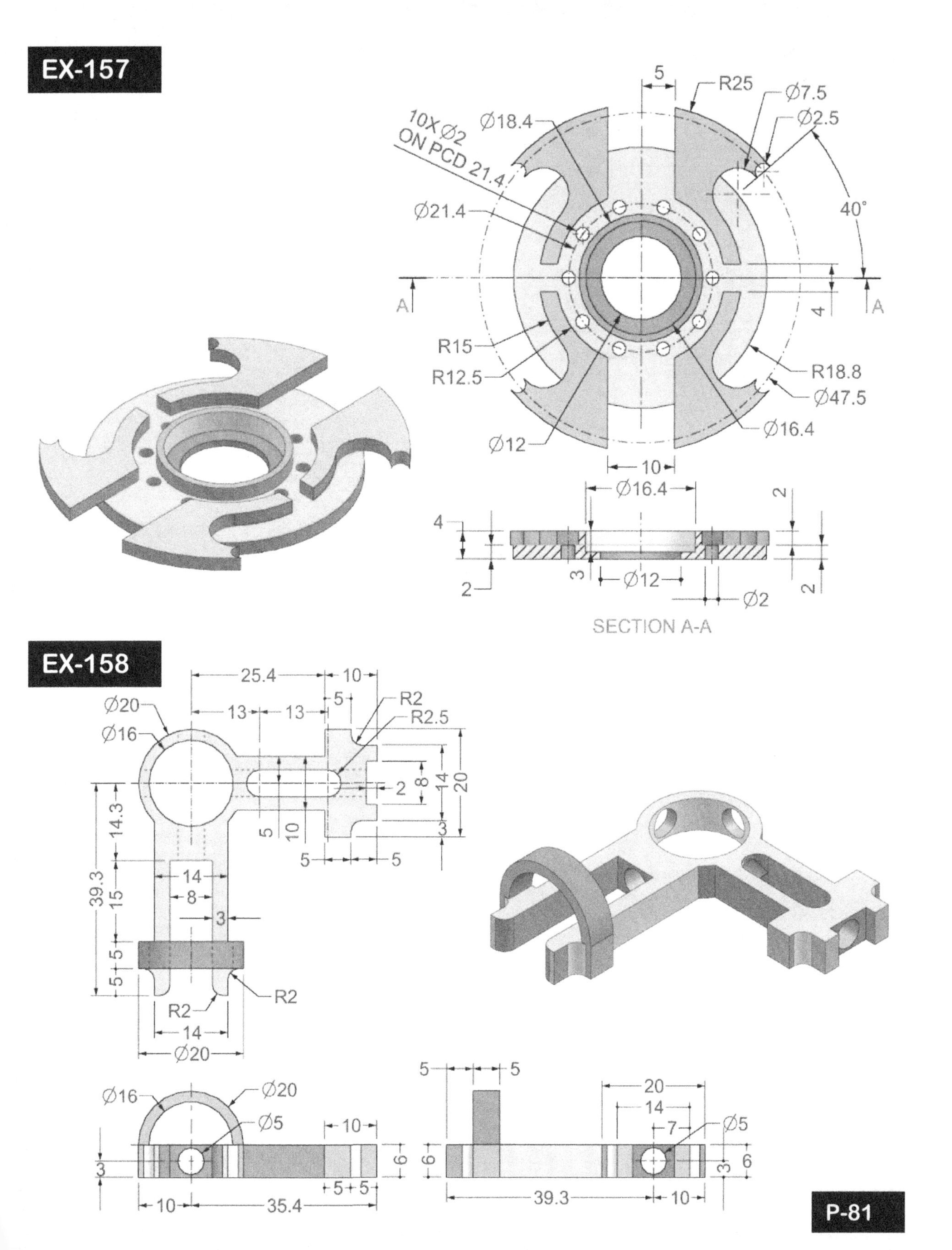
EX-157
5
R25
Ø7.5
Ø2.5
10X Ø2
ON PCD 21.4
Ø18.4
Ø21.4
40°
A
A
4
R15
R12.5
R18.8
Ø47.5
Ø16.4
Ø12
10
Ø16.4
2
4
2
3
Ø12
Ø2
2
SECTION A-A
EX-158
25.4
10
Ø20
13
13
5
R2
R2.5
Ø16
2
8
14
20
14.3
5
10
3
5
5
39.3
14
15
8
3
5
5
R2
R2
14
Ø20
Ø16
Ø20
Ø5
10
3
6
10
35.4
5
5
5
5
20
14
7
Ø5
6
6
3
39.3
10
P-81

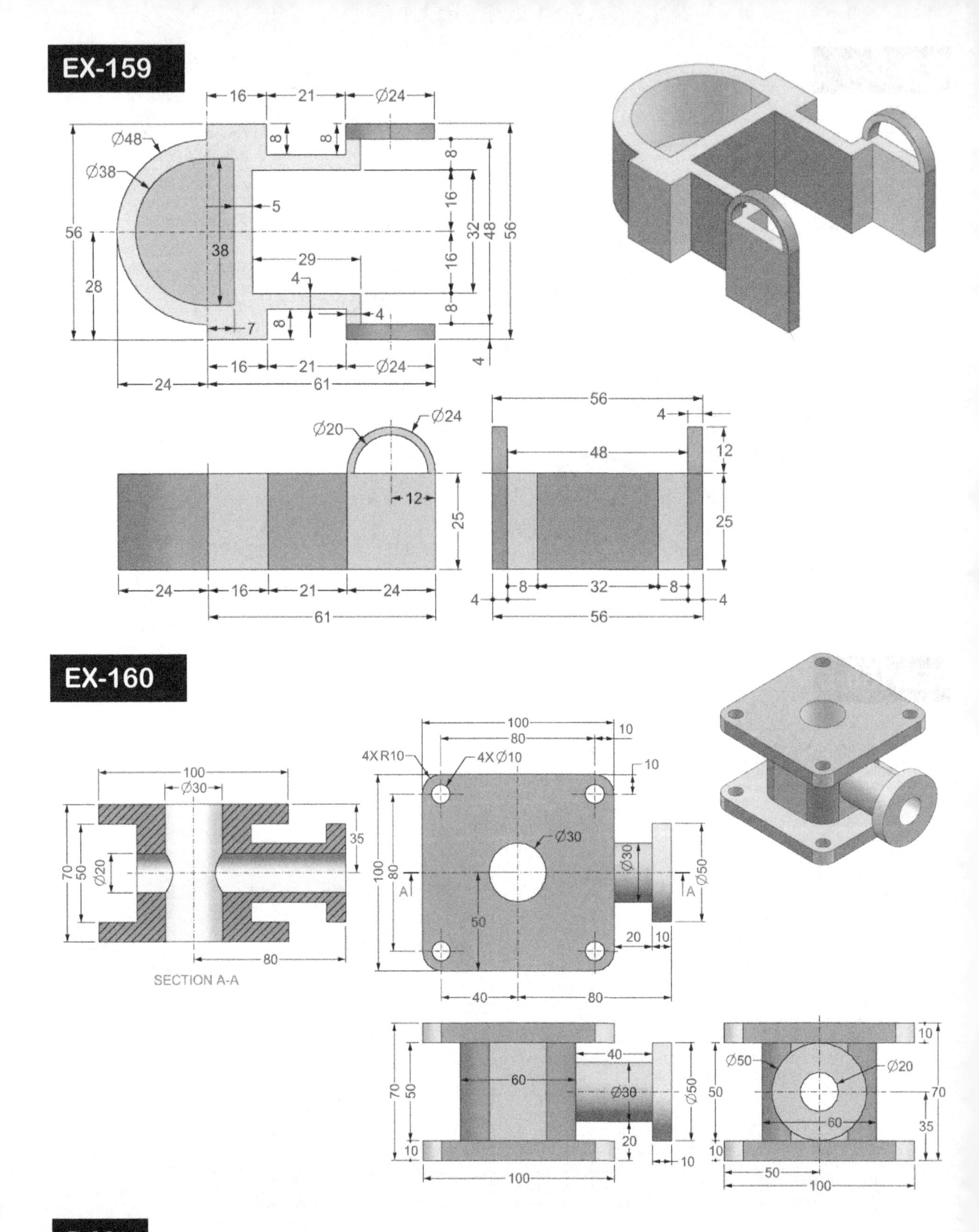
EX-159
Ø48
Ø38
Ø24
Ø20
EX-160
4X R10
4X Ø10
Ø30
Ø50
Ø20
SECTION A-A

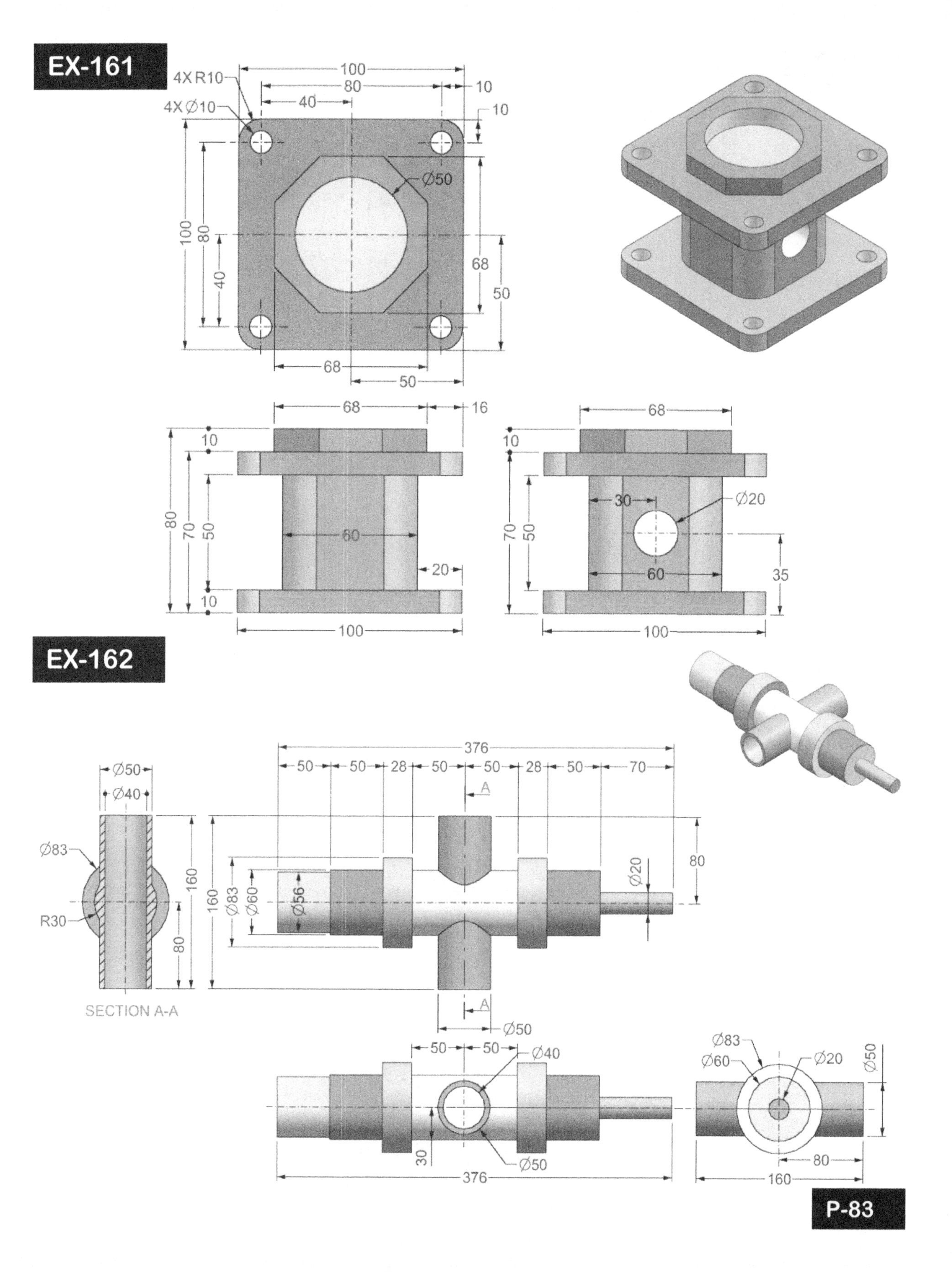
EX-161
4X R10
4X Ø10
100
80
40
10
10
Ø50
100
80
40
68
50
68
50
68
16
10
80
70
50
60
20
10
100
68
10
30
Ø20
70
50
60
35
100
EX-162
Ø50
Ø40
Ø83
R30
160
80
SECTION A-A
376
50
50
28
50
50
28
50
70
A
160
Ø83
Ø60
Ø56
Ø20
80
A
Ø50
50
50
Ø40
30
Ø50
376
Ø83
Ø60
Ø20
Ø50
80
160
P-83

EX-163

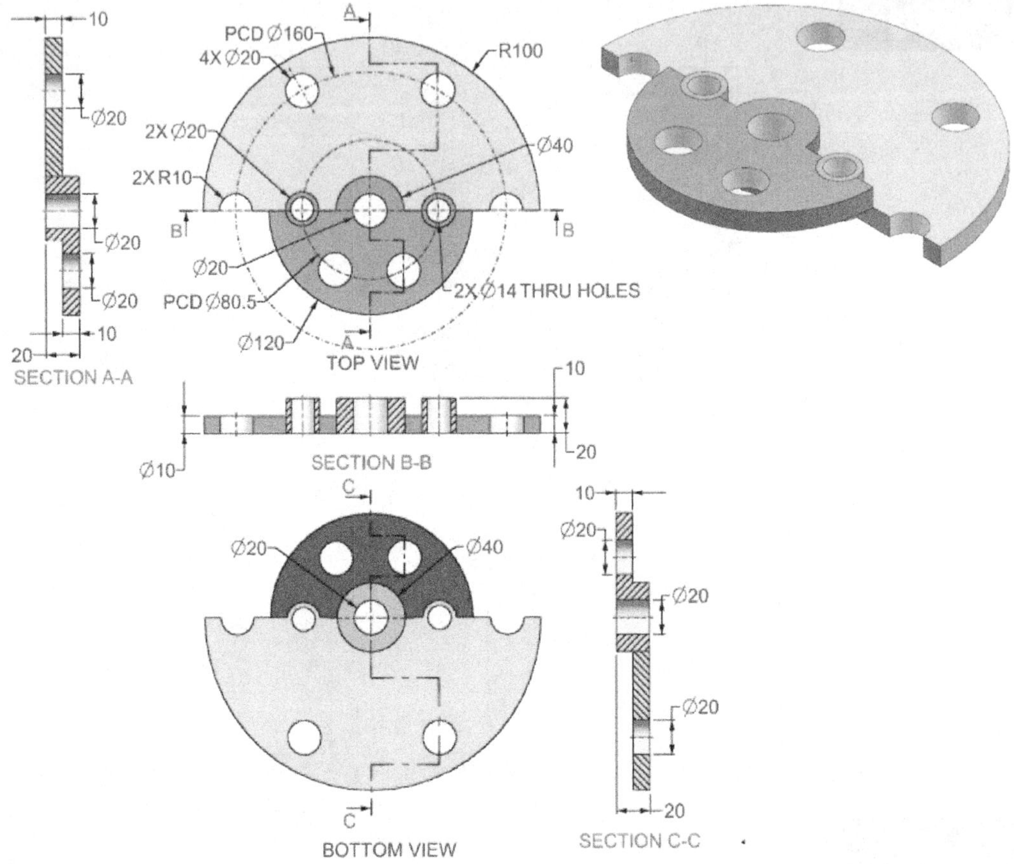

10
Ø20
Ø20
Ø20
10
20
SECTION A-A

A
PCD Ø160
4X Ø20
R100
2X Ø20
Ø40
2X R10
B
B
Ø20
2X Ø14 THRU HOLES
PCD Ø80.5
Ø120
A
TOP VIEW

10
Ø10
20
SECTION B-B

C
Ø20
Ø40
C
BOTTOM VIEW

10
Ø20
Ø20
Ø20
20
SECTION C-C

EX-164

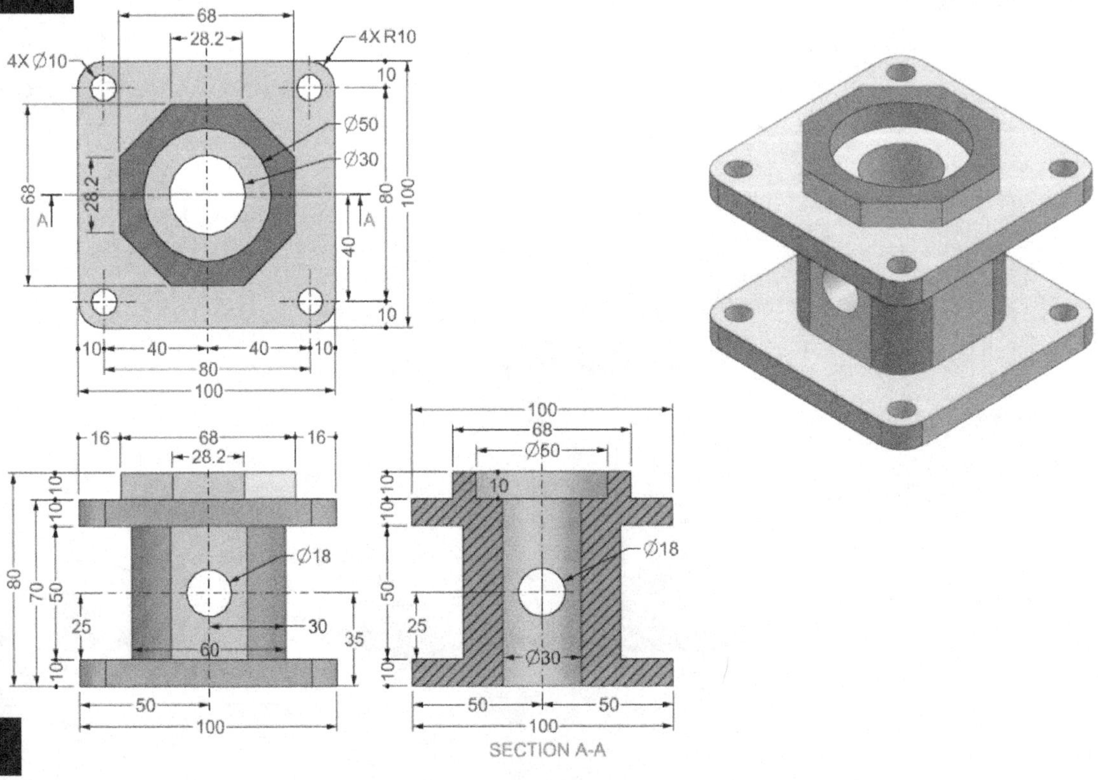

68
28.2
4X R10
4X Ø10
10
Ø50
Ø30
68
28.2
A
A
80
100
40
10
10
40
40
10
80
100

16
68
16
28.2
10
10
Ø18
80
70
50
25
30
35
60
10
50
100

100
68
Ø50
10
10
10
Ø18
50
25
Ø30
10
50
50
100
SECTION A-A

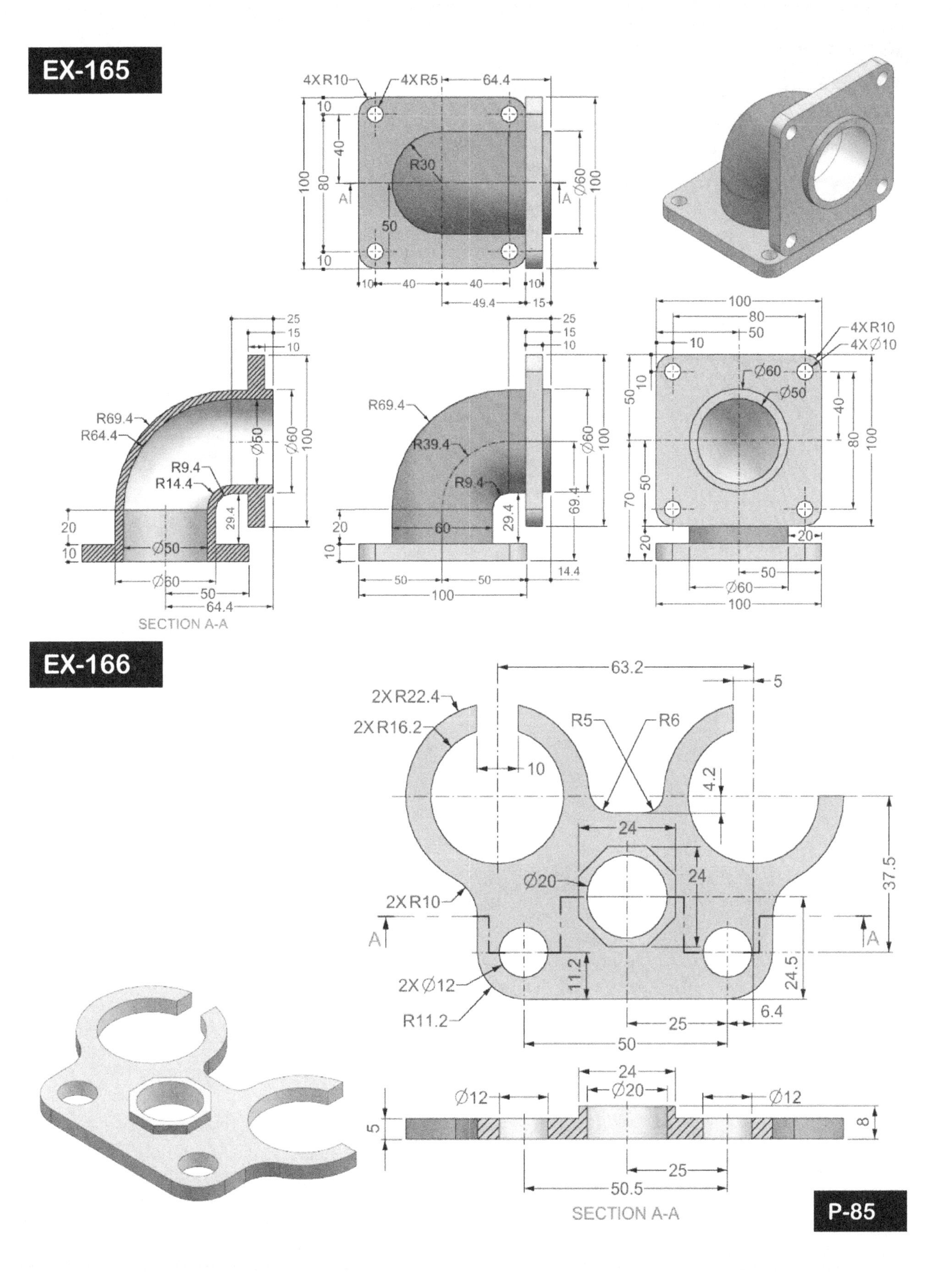
EX-165
4X R10
4X R5
64.4
R30
40
50
80
100
Ø60
A
10
49.4
15
R69.4
R64.4
R9.4
R14.4
Ø50
29.4
20
SECTION A-A
R39.4
60
69.4
14.4
4X Ø10
70
EX-166
63.2
5
2X R22.4
2X R16.2
R5
R6
4.2
24
Ø20
2X R10
37.5
2X Ø12
11.2
24.5
R11.2
6.4
25
Ø12
8
50.5
SECTION A-A
P-85

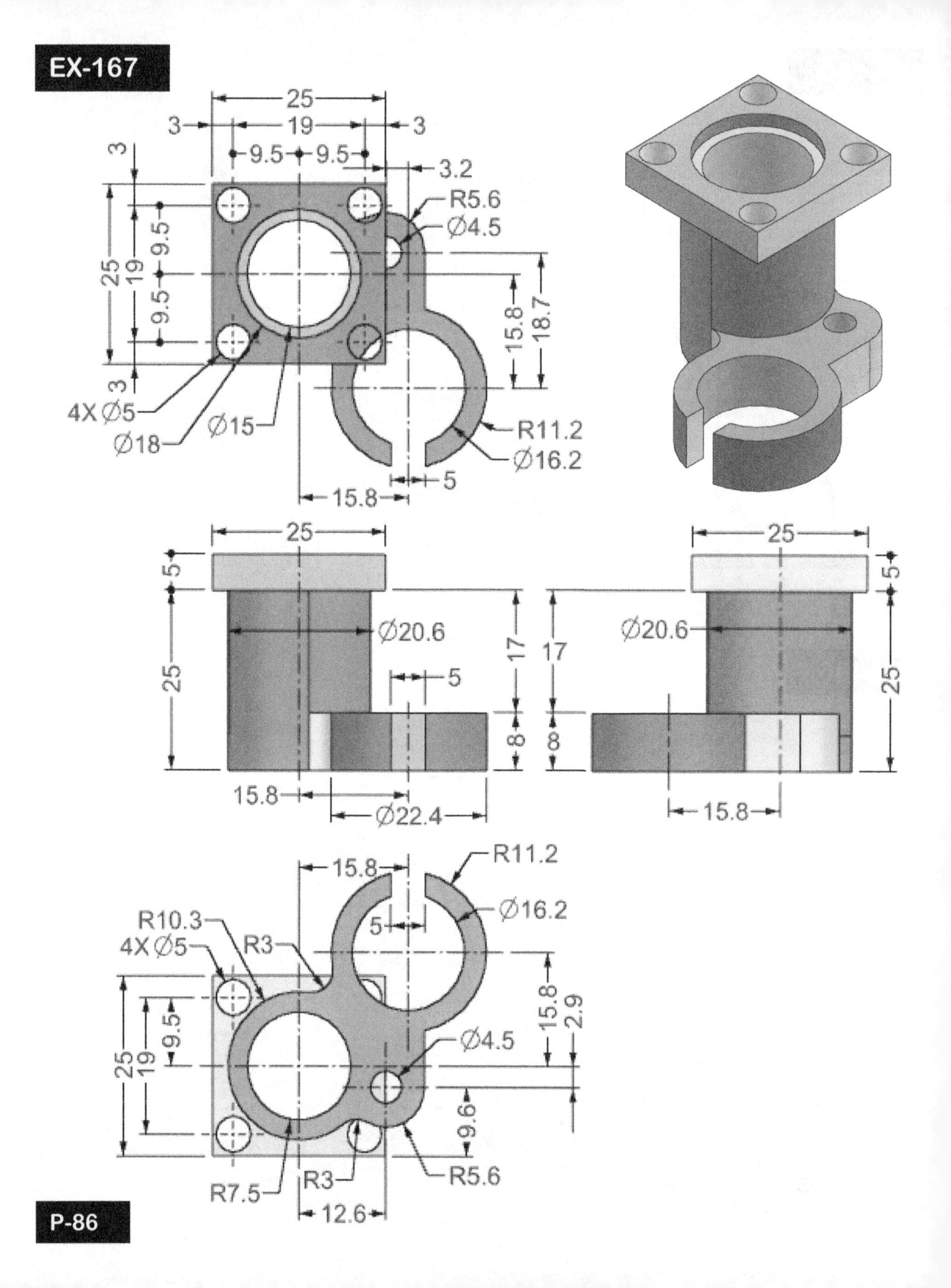
EX-167
25
3
19
3
9.5
9.5
3.2
R5.6
Ø4.5
25
19
9.5
9.5
3
3
15.8
18.7
4X Ø5
Ø18
Ø15
R11.2
Ø16.2
5
15.8
25
5
25
Ø20.6
17
5
8
15.8
Ø22.4
25
5
Ø20.6
17
8
25
15.8
15.8
R11.2
Ø16.2
R10.3
4X Ø5
R3
5
25
19
9.5
15.8
2.9
Ø4.5
9.6
R7.5
R3
R5.6
12.6

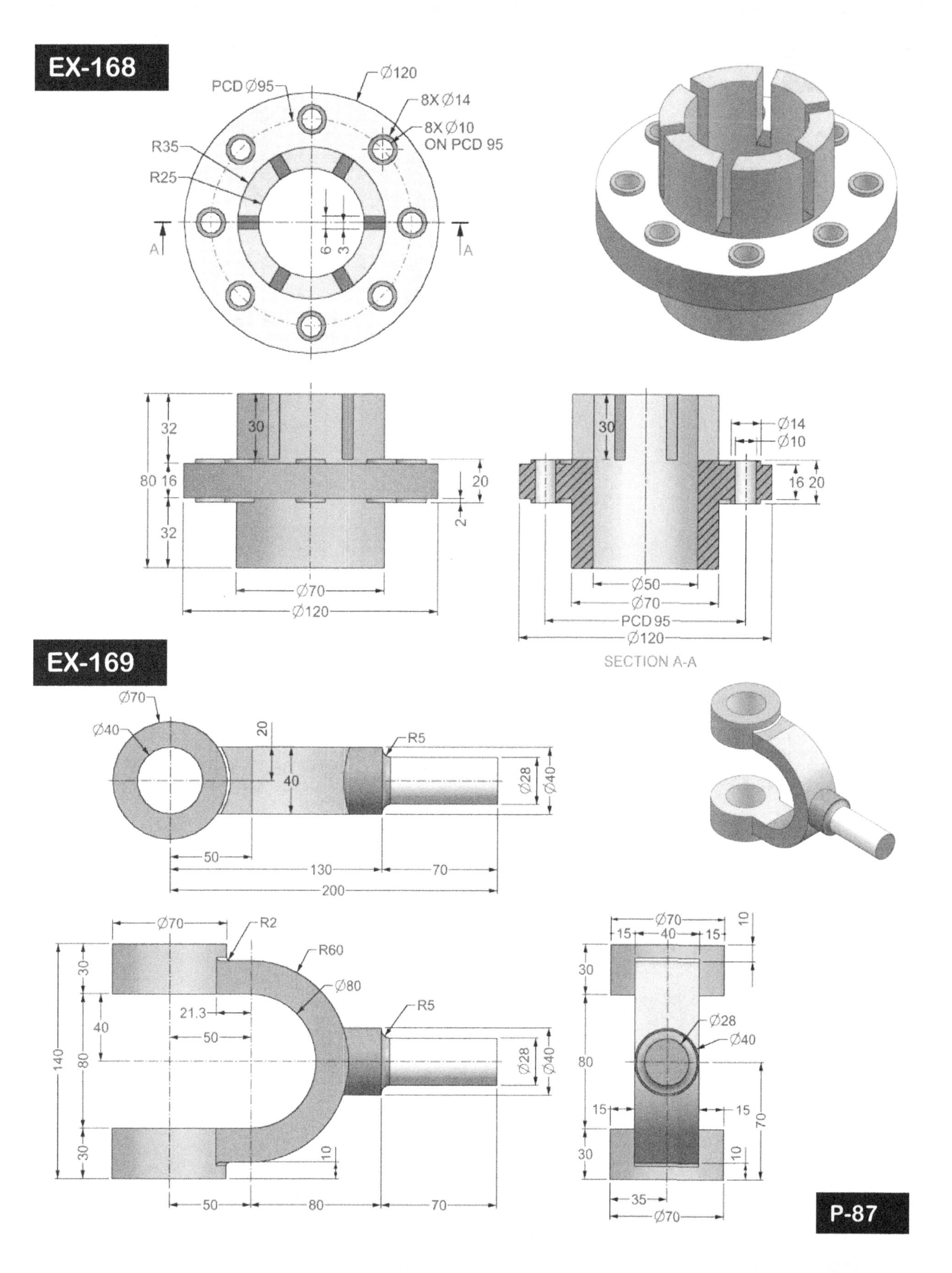
EX-168
PCD Ø95
Ø120
8X Ø14
8X Ø10
ON PCD 95
R35
R25
A
A
6
3
32
30
80 16
32
20
2
Ø70
Ø120
30
Ø14
Ø10
16 20
Ø50
Ø70
PCD 95
Ø120
SECTION A-A
EX-169
Ø70
Ø40
20
40
R5
Ø28
Ø40
50
130
70
200
Ø70
R2
R60
Ø80
21.3
50
40
R5
140
80
30
30
Ø28
Ø40
10
50
80
70
Ø70
15
40
15
10
30
Ø28
Ø40
80
15
15
70
30
10
35
Ø70
P-87

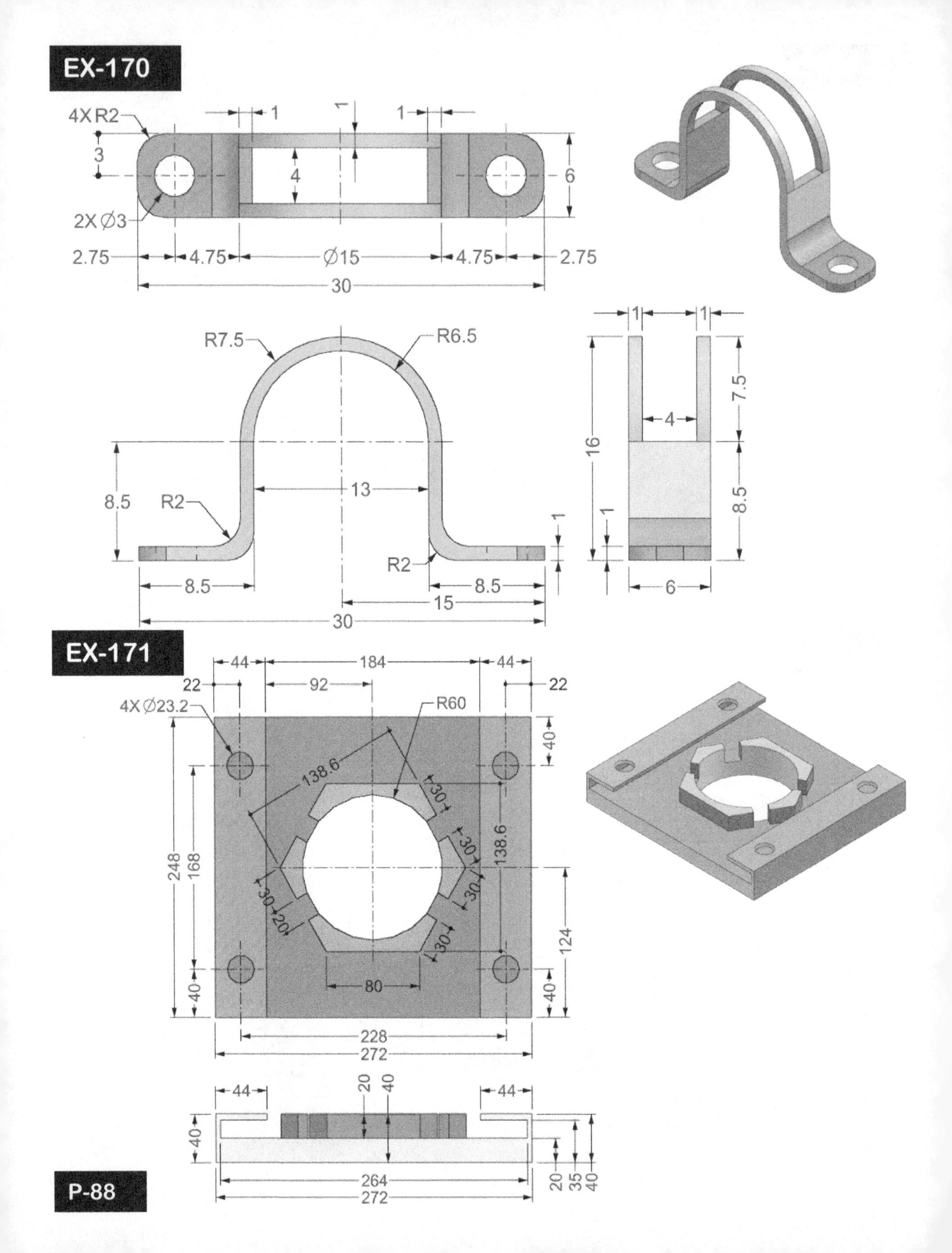

EX-170
4X R2
2X Ø3
3
1
1
1
4
6
2.75
4.75
Ø15
4.75
2.75
30
R7.5
R6.5
8.5
R2
13
R2
1
8.5
8.5
15
30
1
1
4
7.5
16
8.5
1
6
EX-171
44
184
44
22
92
22
4X Ø23.2
R60
40
138.6
30
30
138.6
248
168
30
30
20
30
124
80
40
40
228
272
44
20
40
44
40
264
272
20
35
40
P-88

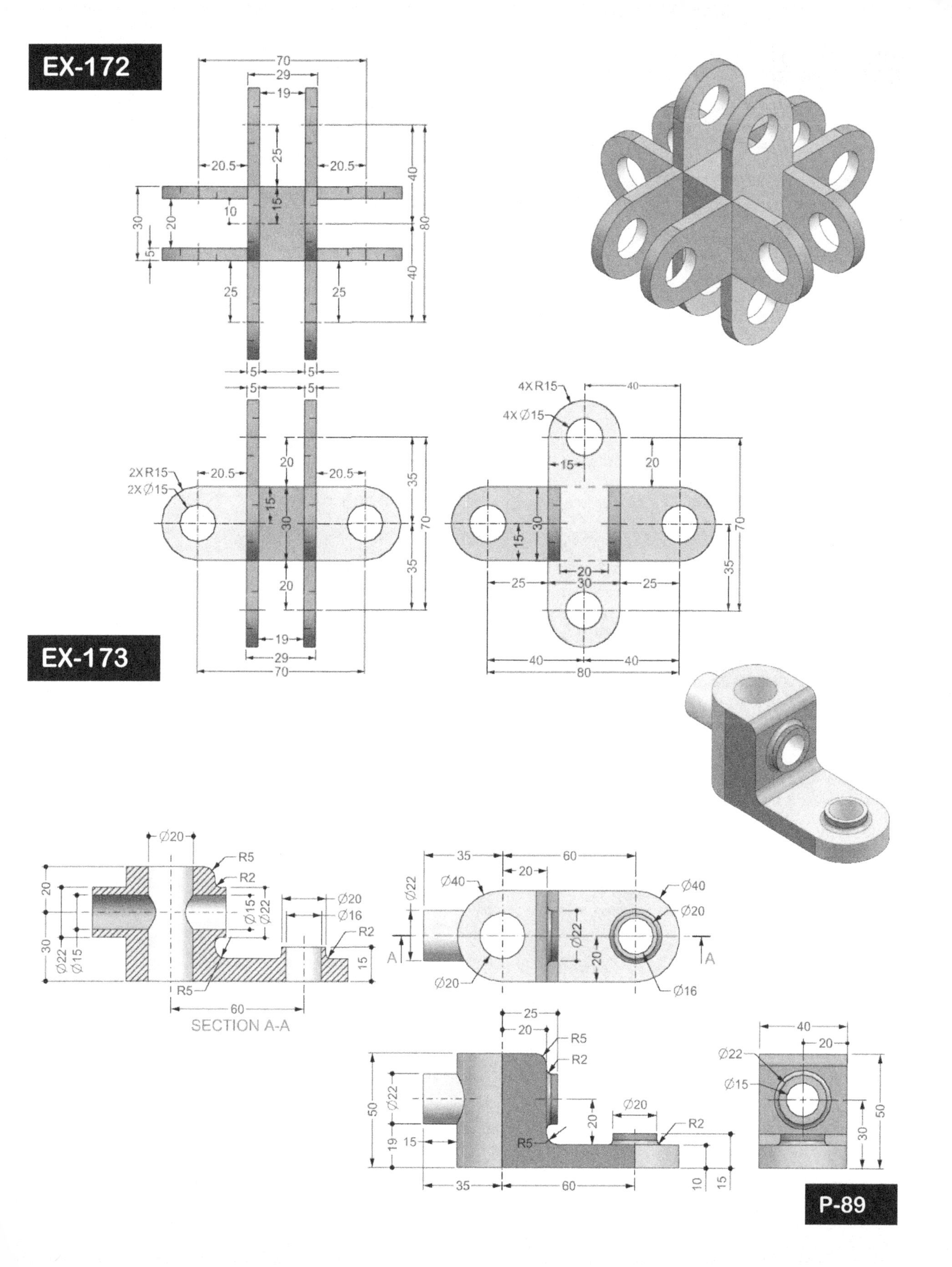
EX-172
EX-173
4X R15
4X Ø15
2X R15
2X Ø15
SECTION A-A
A
A
P-89

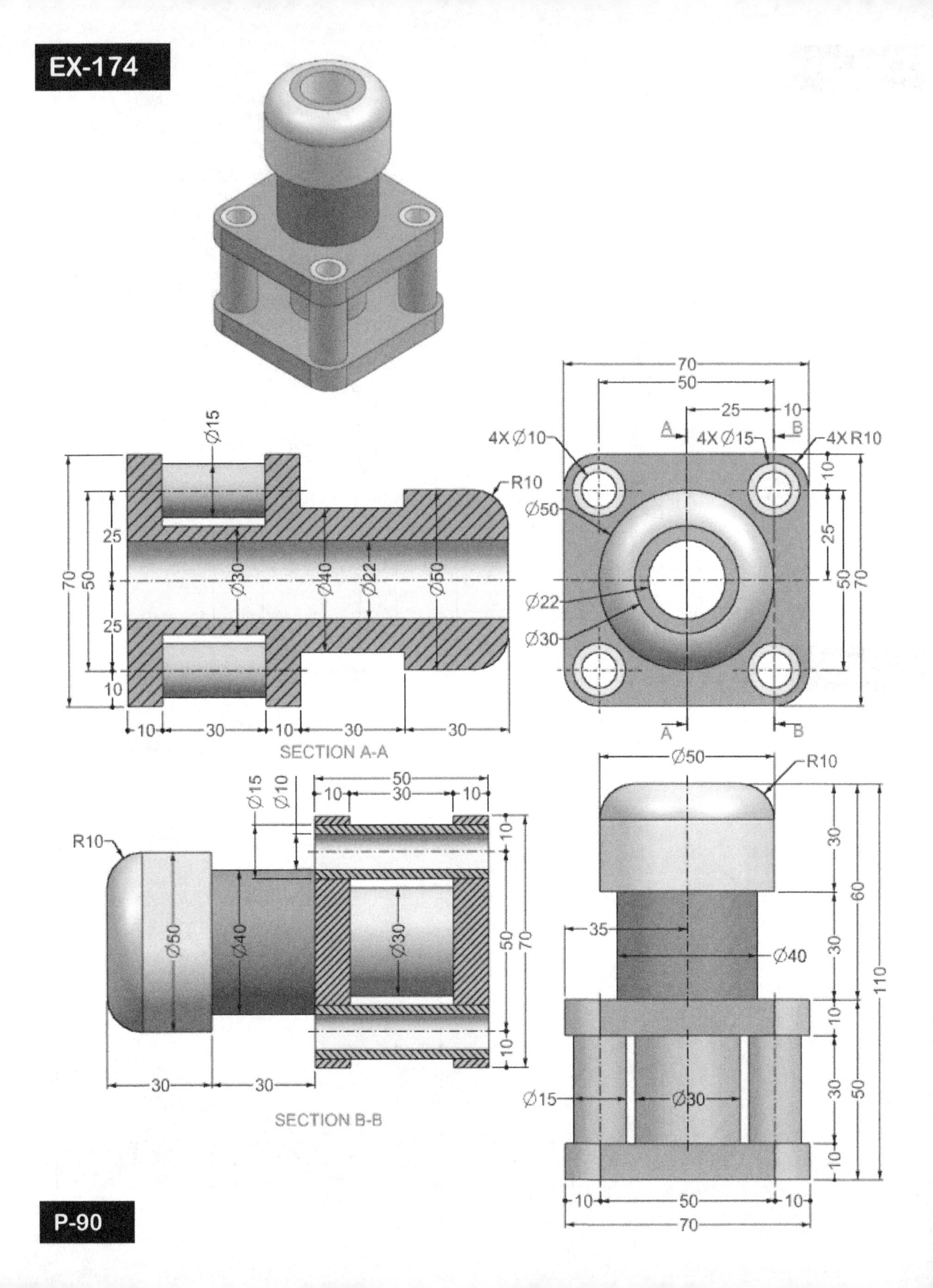
EX-174
70
50
25
10
4X Ø10
A
4X Ø15
B
4X R10
Ø15
R10
Ø50
25
Ø30
Ø40
Ø22
Ø50
Ø22
Ø30
10
30
SECTION A-A
Ø50
R10
Ø15
Ø10
50
10
30
10
R10
Ø50
Ø40
Ø30
50
70
SECTION B-B
35
Ø40
60
110
Ø15
Ø30
P-90

EX-175

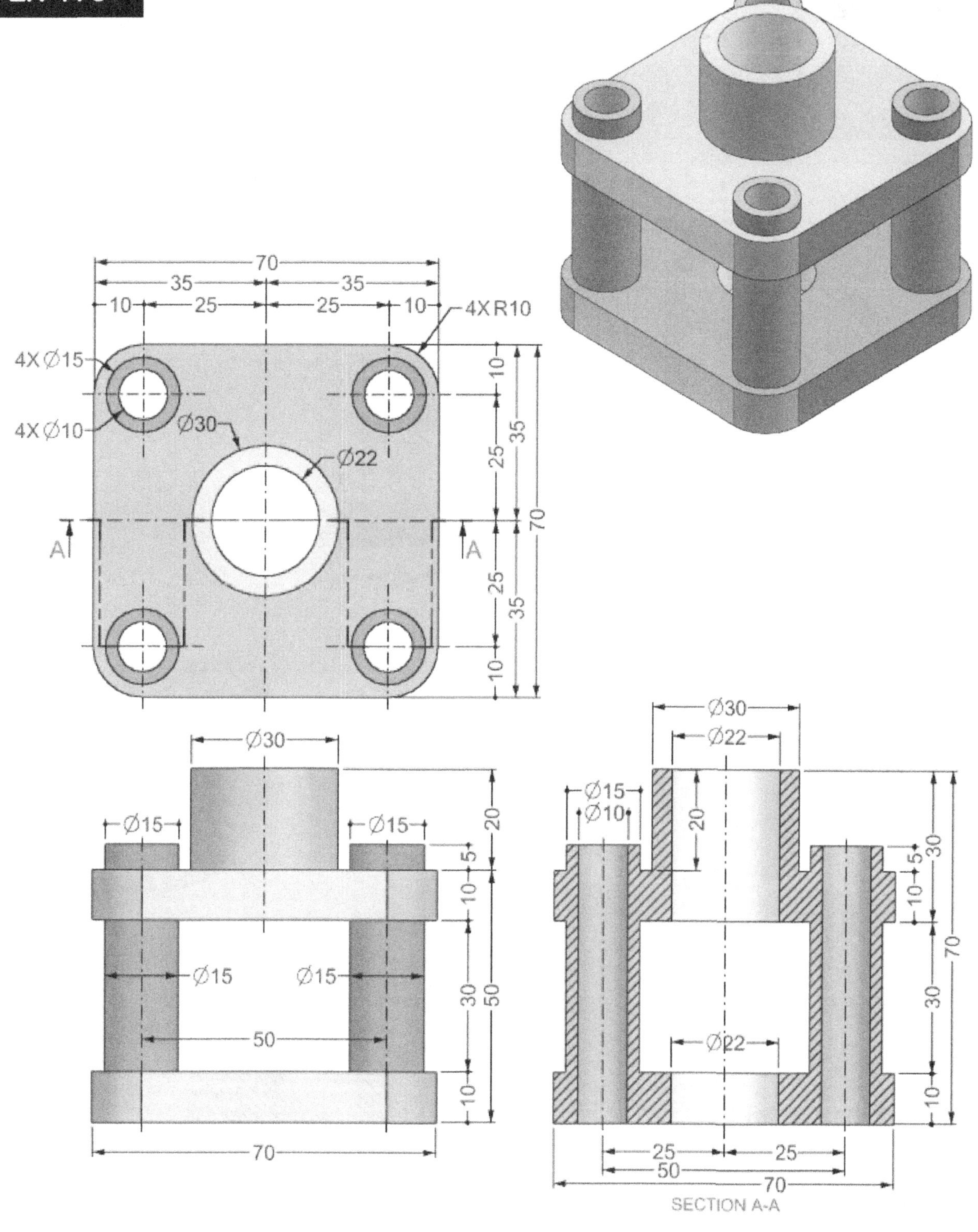

70
35
35
10
25
25
10
4X R10
4X ∅15
4X ∅10
∅30
∅22
A
A
10
25
35
70
25
35
10
∅30
∅15
∅15
20
5
10
∅15
∅15
30
50
50
10
70
∅30
∅22
∅15
∅10
20
5
30
10
70
30
∅22
10
25
25
50
70
SECTION A-A

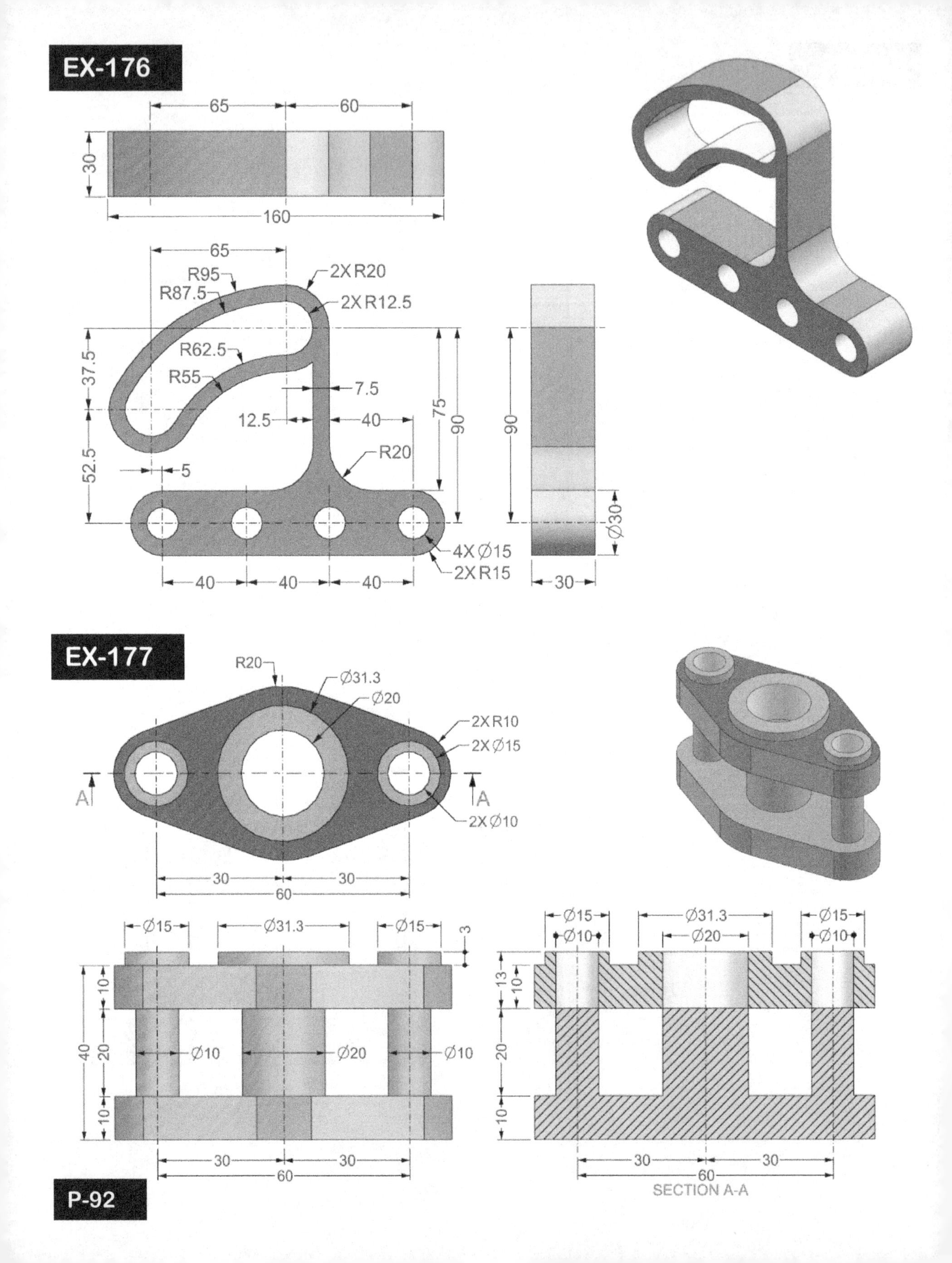

EX-176
65
60
30
160
65
R95
R87.5
2X R20
2X R12.5
R62.5
R55
37.5
52.5
7.5
12.5
40
75
90
R20
5
4X Ø15
2X R15
40
40
40
90
Ø30
30
EX-177
R20
Ø31.3
Ø20
2X R10
2X Ø15
2X Ø10
A
A
30
30
60
Ø15
Ø31.3
Ø15
3
10
40
20
10
Ø10
Ø20
Ø10
30
30
60
Ø15
Ø10
Ø31.3
Ø20
Ø15
Ø10
13
10
20
10
30
30
60
SECTION A-A

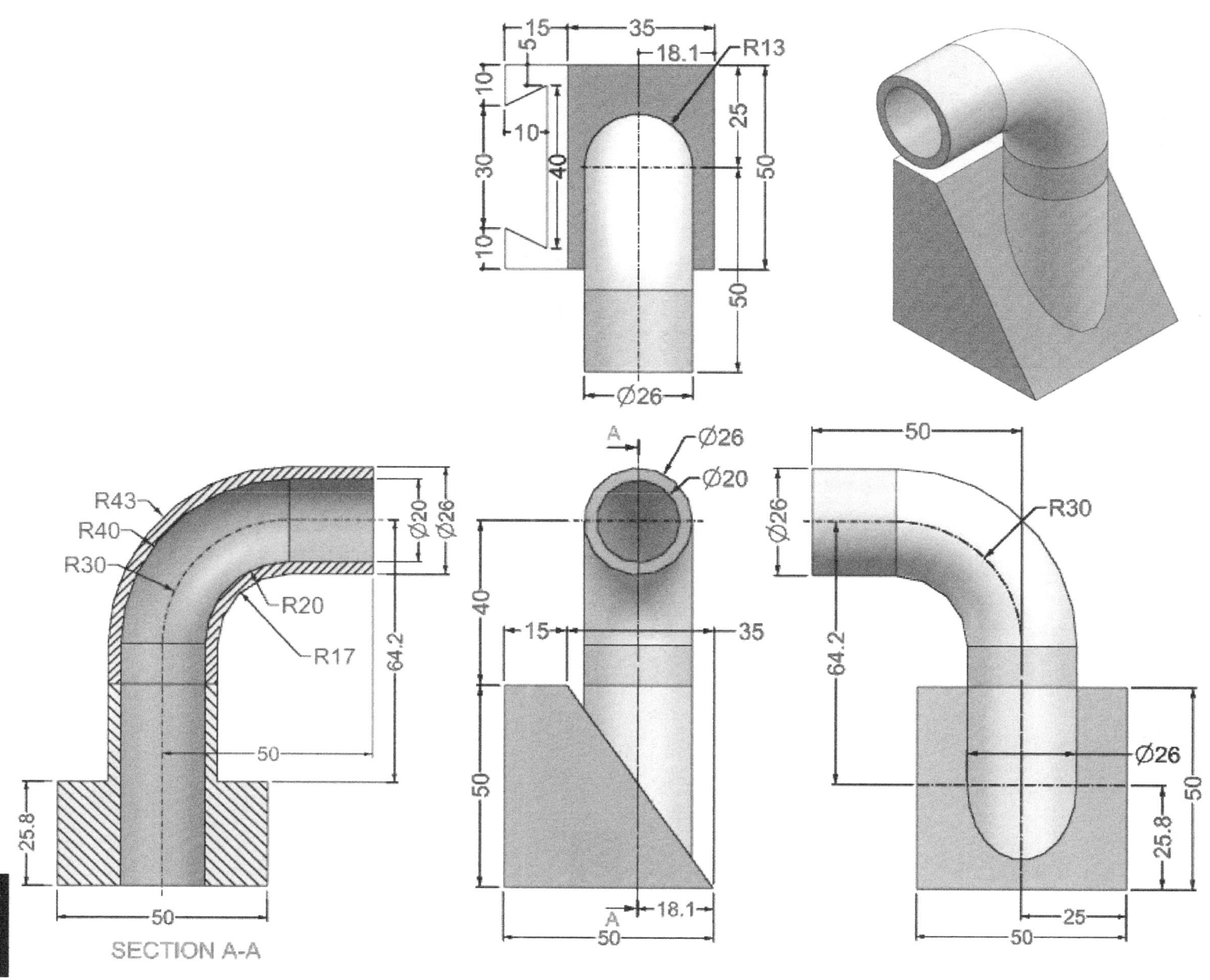
15
35
5
18.1
R13
10
10
30
40
25
50
10
50
Ø26
A
Ø26
Ø20
50
R43
R40
R30
Ø20
Ø26
R20
R17
64.2
50
25.8
50
SECTION A-A
40
15
35
50
A
18.1
50
Ø26
R30
64.2
Ø26
25.8
50
25
50

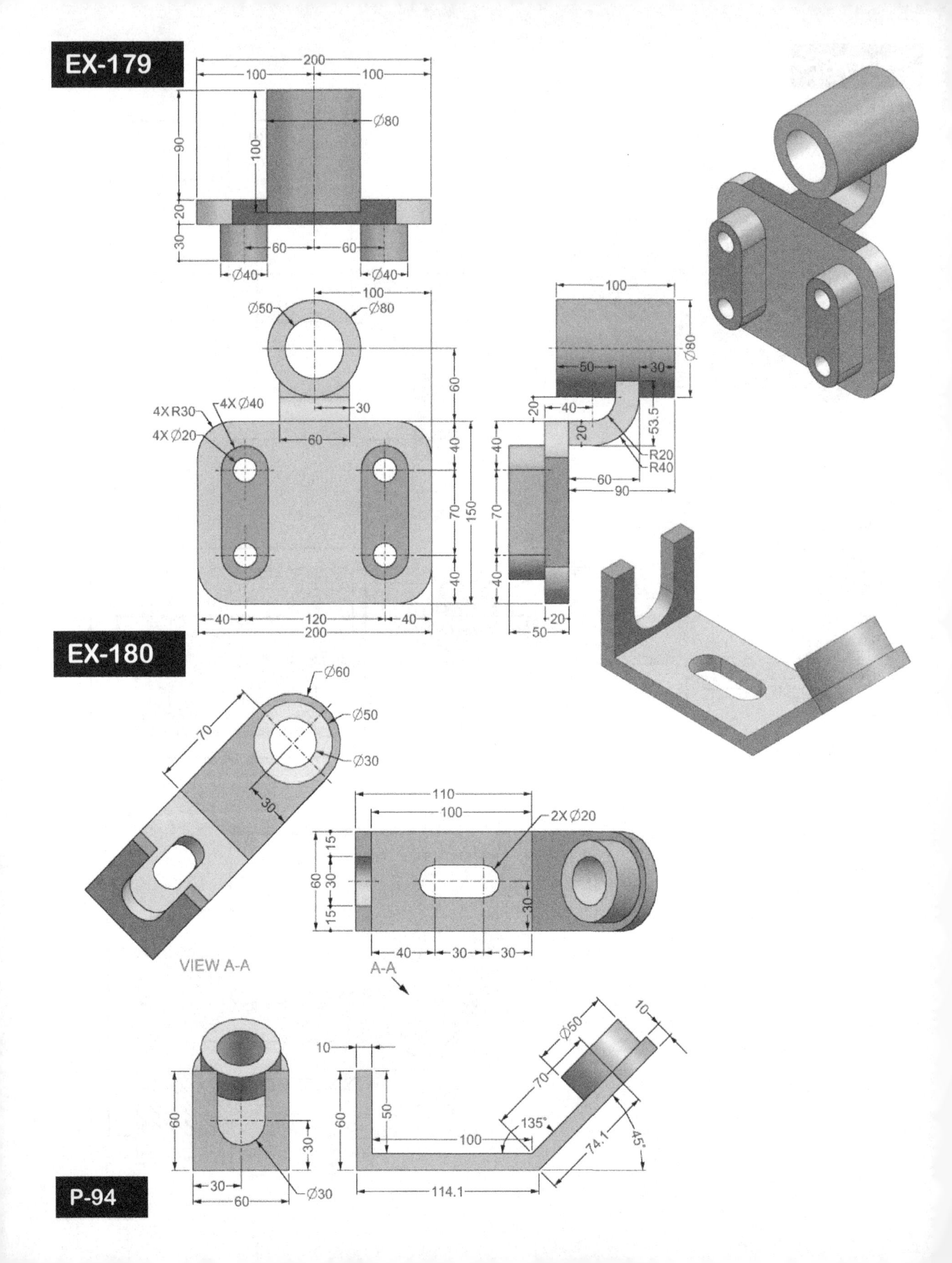
EX-179
200
100
100
Ø80
90
100
20
30
60
60
Ø40
Ø40
100
Ø50
Ø80
60
4X R30
4X Ø40
30
4X Ø20
60
40
70
150
40
40
120
40
200
100
50
30
Ø80
20
40
53.5
20
R20
R40
40
60
90
70
40
20
50
EX-180
Ø60
70
Ø50
Ø30
30
110
100
2X Ø20
15
60
30
15
30
40
30
30
VIEW A-A
A-A
10
Ø50
10
70
60
50
135°
100
74.1
45°
60
30
30
60
Ø30
114.1
P-94

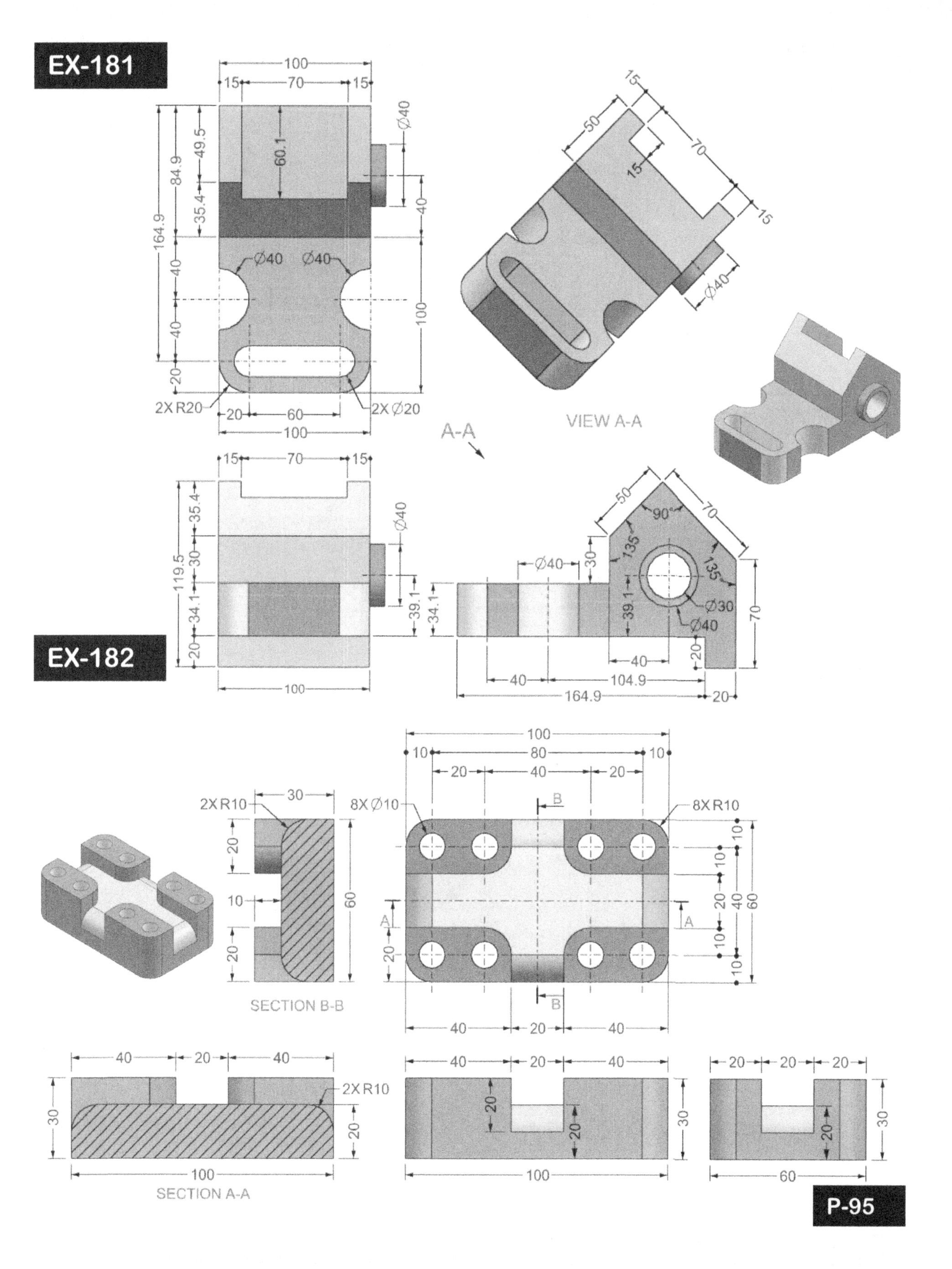
EX-181
100
15
70
15
Ø40
49.5
60.1
84.9
35.4
164.9
40
Ø40
Ø40
40
100
40
20
2X R20
2X Ø20
20
60
100
VIEW A-A
A-A
EX-182
15
70
15
35.4
Ø40
119.5
30
34.1
39.1
20
100
50
90°
70
135°
135°
30
Ø40
Ø30
Ø40
39.1
34.1
70
40
20
40
104.9
164.9
20
100
10
80
10
20
40
20
2X R10
30
8X Ø10
B
8X R10
20
10
60
10
20
40
60
A
A
20
10
SECTION B-B
40
20
40
40
20
40
2X R10
30
20
100
SECTION A-A
40
20
40
20
20
30
100
20
20
20
20
30
60
P-95

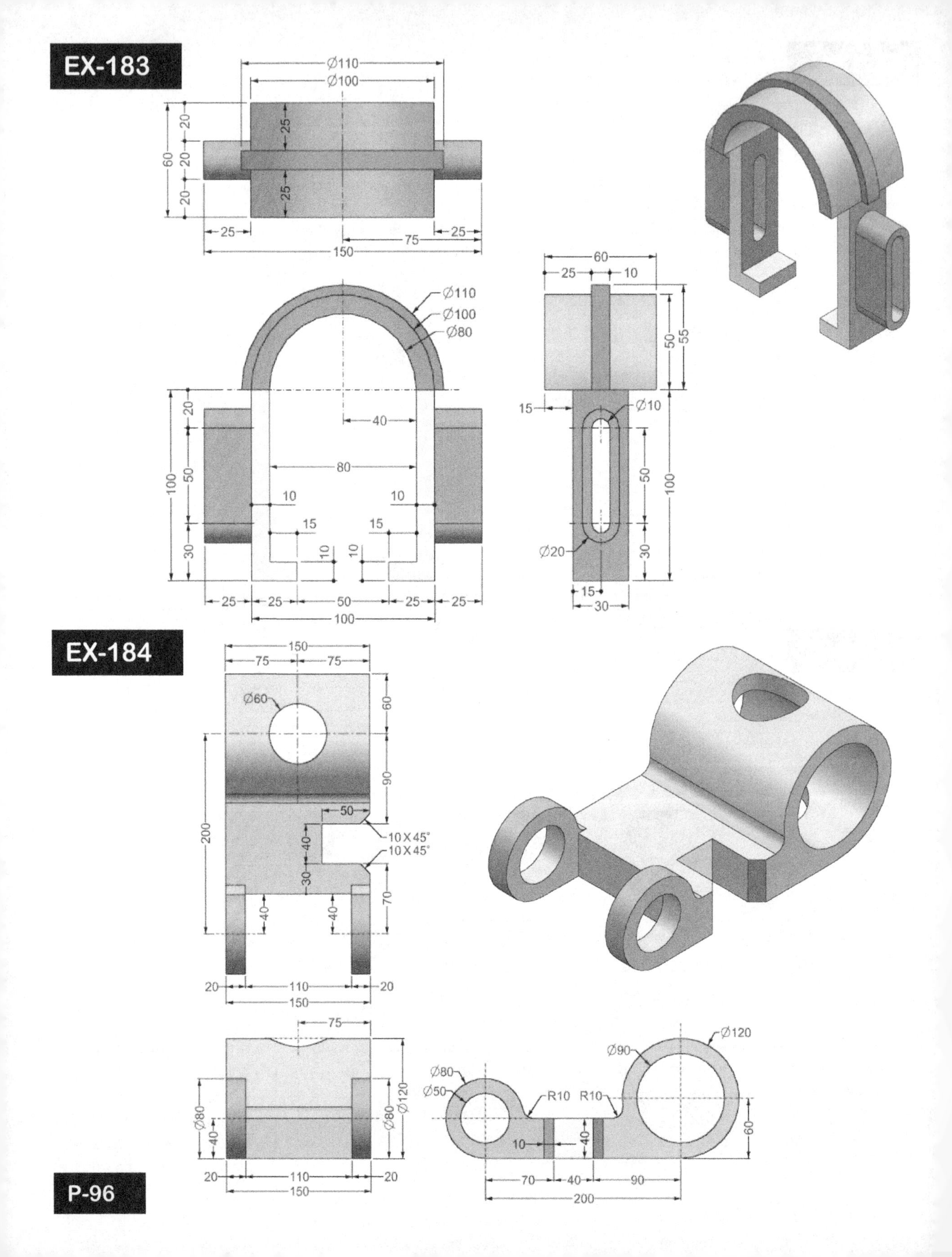
EX-183
Ø110
Ø100
20
20
20
60
25
25
25
75
25
150
Ø110
Ø100
Ø80
60
25
10
50
55
20
40
80
100
50
10
10
15
15
10
10
30
15
Ø10
Ø20
50
100
30
15
30
25
25
50
25
25
100
EX-184
150
75
75
Ø60
60
90
50
10 X 45°
10 X 45°
40
30
200
40
40
70
20
110
20
150
75
Ø80
40
Ø80
Ø120
20
110
20
150
Ø120
Ø90
Ø80
Ø50
R10
R10
10
40
60
70
40
90
200
P-96

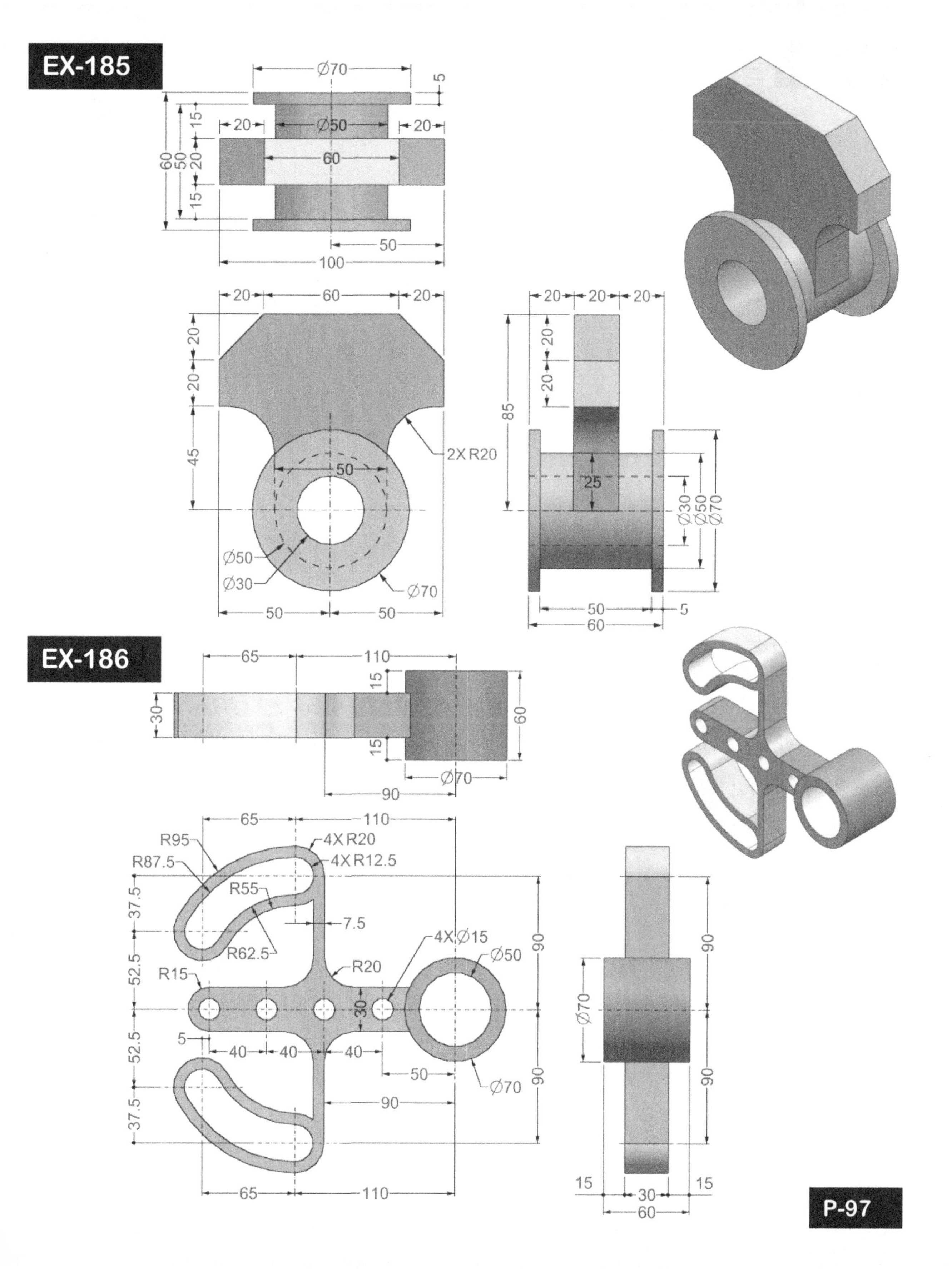
EX-185
Ø70
5
15
20
20
60
50
60
15
Ø50
20
20
50
100
20
60
20
20
20
45
50
2X R20
Ø50
Ø30
Ø70
50
50
20
20
20
85
20
20
25
Ø30
Ø50
Ø70
50
5
60
EX-186
65
110
15
30
60
15
Ø70
90
65
110
R95
4X R20
R87.5
4X R12.5
R55
37.5
7.5
R62.5
4X Ø15
Ø50
52.5
R15
R20
30
90
5
40
40
40
52.5
50
Ø70
90
90
37.5
65
110
90
Ø70
90
15
30
15
60
P-97

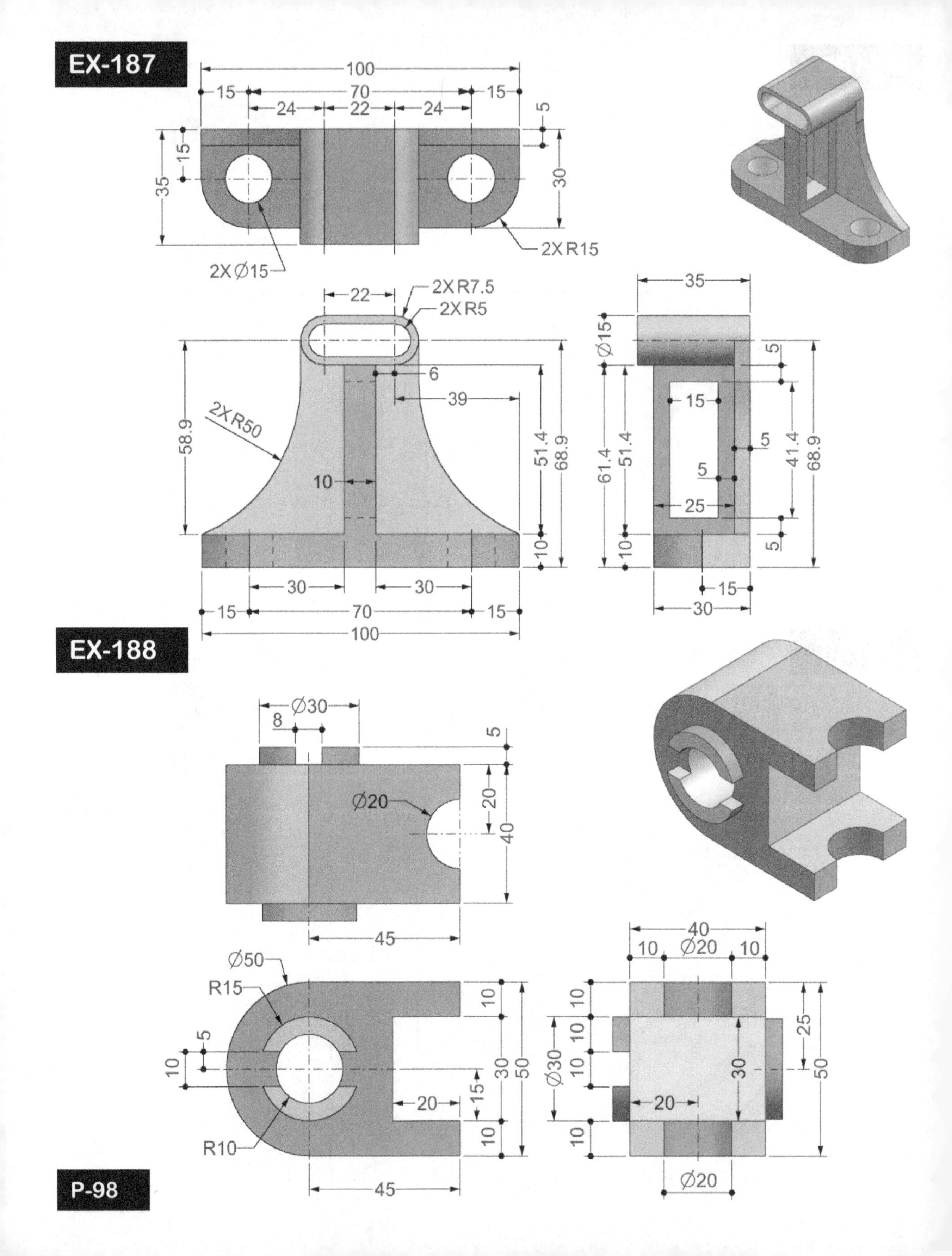
EX-187
100
15
70
15
24
22
24
5
15
35
30
2X R15
2X Ø15
22
2X R7.5
2X R5
35
Ø15
5
6
39
15
2X R50
58.9
51.4
68.9
61.4
51.4
5
41.4
68.9
5
10
25
10
10
5
30
30
15
15
70
15
100
30
EX-188
Ø30
8
5
Ø20
20
40
45
40
10
Ø20
10
Ø50
R15
10
5
10
10
10
30
50
Ø30
10
25
30
50
20
15
20
10
10
R10
45
Ø20
P-98

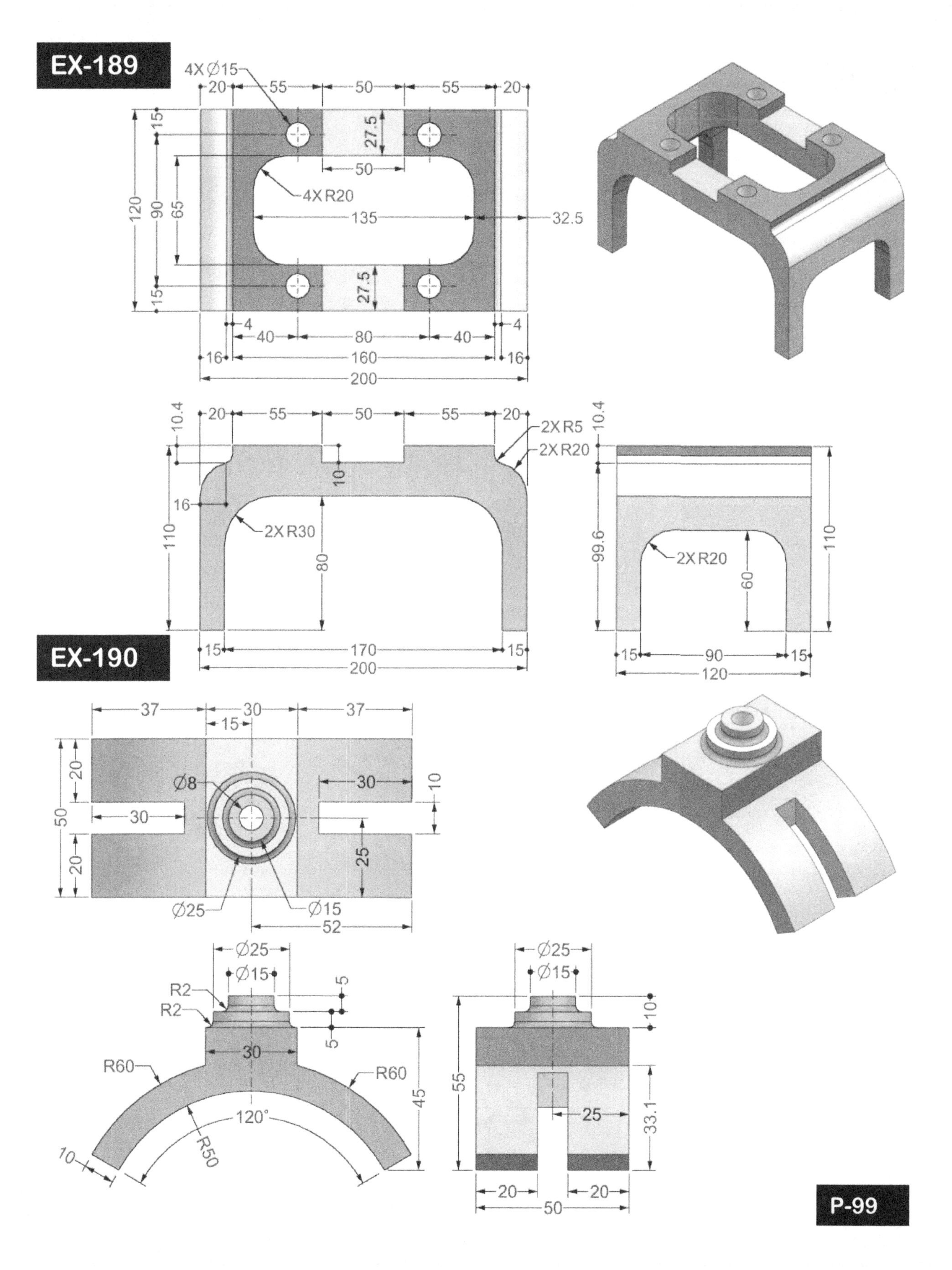
EX-189
4X Ø15
20
55
50
55
20
15
27.5
50
4X R20
120
90
65
135
32.5
15
27.5
4
40
80
40
4
16
160
16
200
10.4
20
55
50
55
20
2X R5
2X R20
10.4
10
16
2X R30
110
80
15
170
15
200
99.6
2X R20
60
110
15
90
15
120
EX-190
37
30
15
37
20
Ø8
30
10
50
30
20
25
Ø25
Ø15
52
Ø25
Ø15
R2
R2
5
5
30
R60
R60
45
120°
10
R50
Ø25
Ø15
10
55
25
33.1
20
20
50
P-99

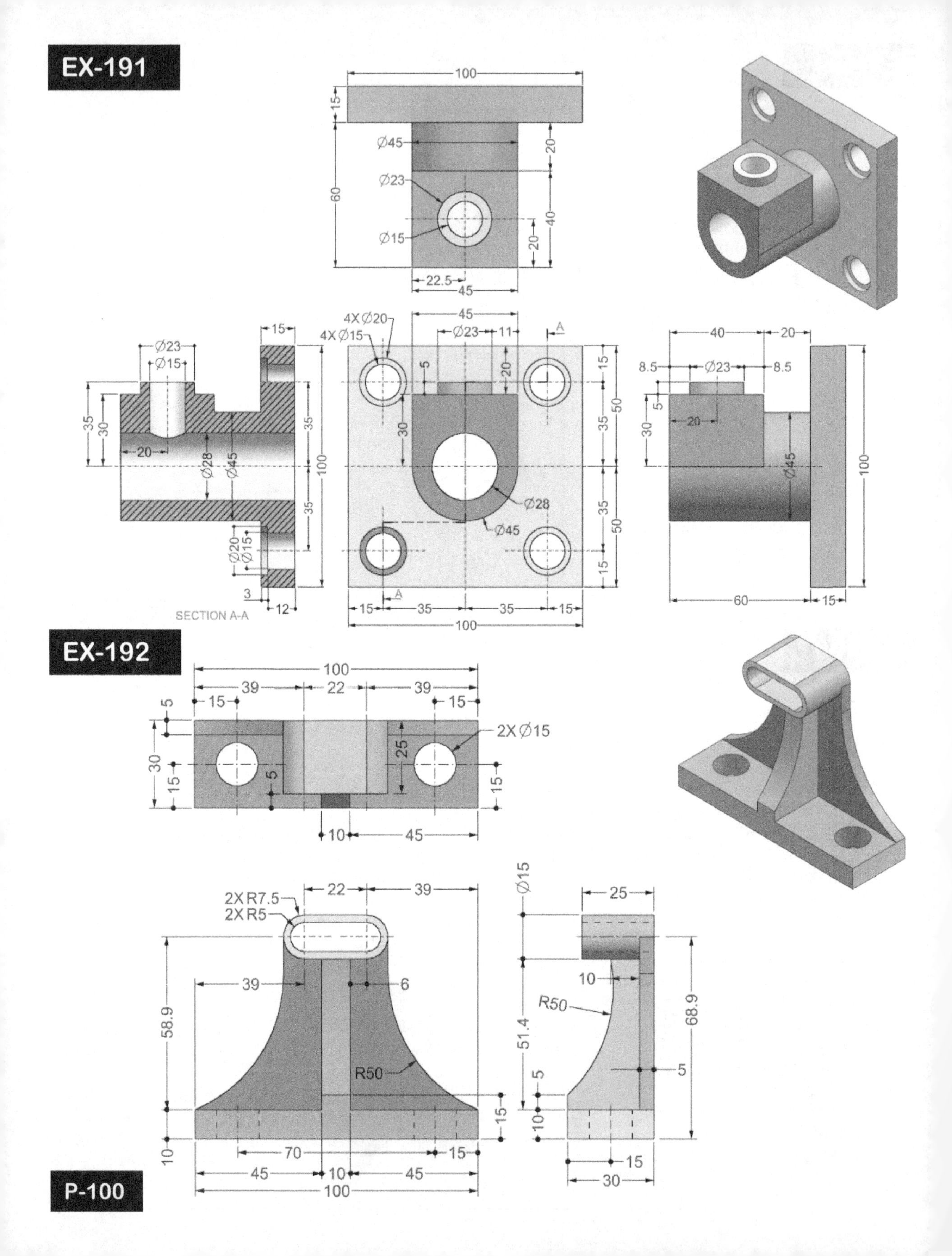

EX-191
100
15
Ø45
20
Ø23
60
40
Ø15
20
22.5
45
4X Ø20
4X Ø15
45
Ø23
11
A
15
Ø23
Ø15
35
30
20
Ø28
Ø45
35
35
100
5
20
30
15
50
35
35
50
15
Ø28
Ø45
Ø20
Ø15
3
12
SECTION A-A
15
35
35
15
100
A
40
20
8.5
Ø23
8.5
5
30
20
Ø45
100
60
15
EX-192
100
39
22
39
15
15
5
2X Ø15
30
25
15
5
15
10
45
2X R7.5
2X R5
22
39
39
6
58.9
R50
10
70
15
45
10
45
100
Ø15
25
10
R50
51.4
68.9
5
5
10
15
30
P-100

EX-193

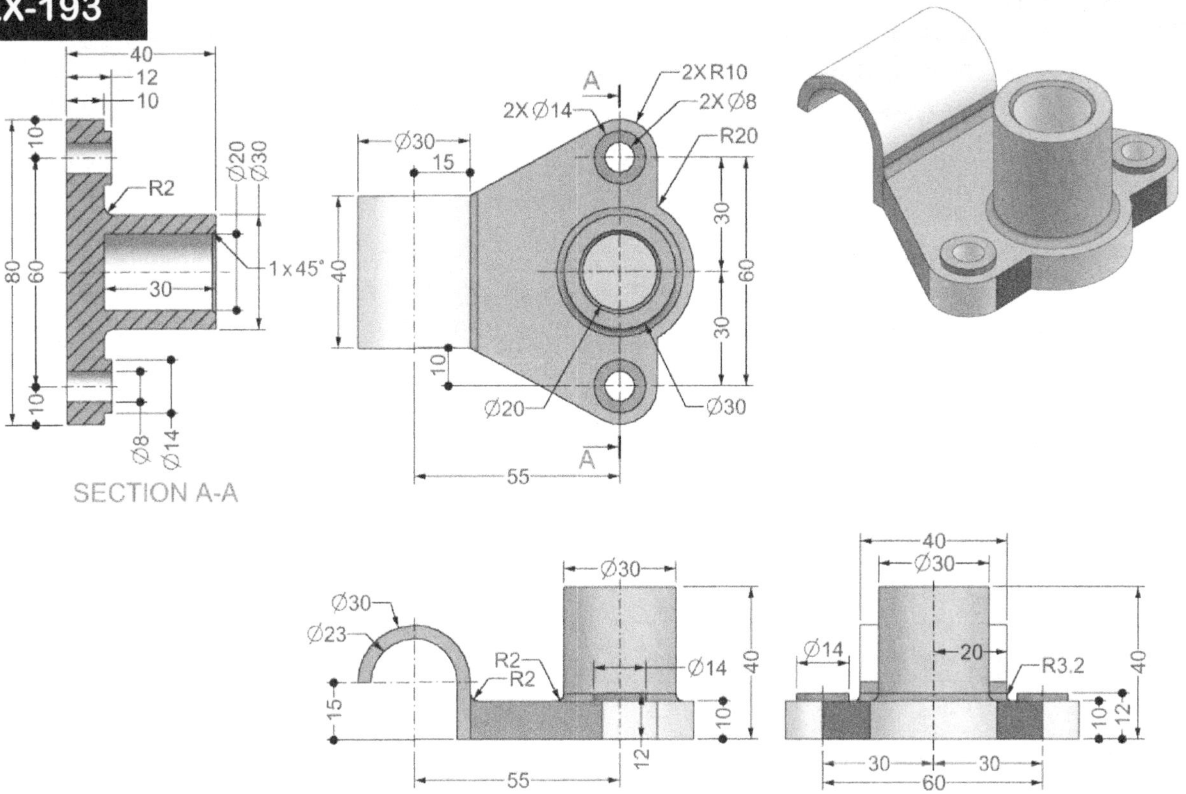

SECTION A-A
2X R10
2X Ø8
2X Ø14
R20
Ø30
Ø20
1 x 45°
R2
R3.2
Ø23

EX-194

4X Ø20
R5
Ø120
130°
ALL HOLES CHAMFER 2MM
2X Ø20
2X Ø50
PCD Ø160
Ø100
BOTTOM VIEW

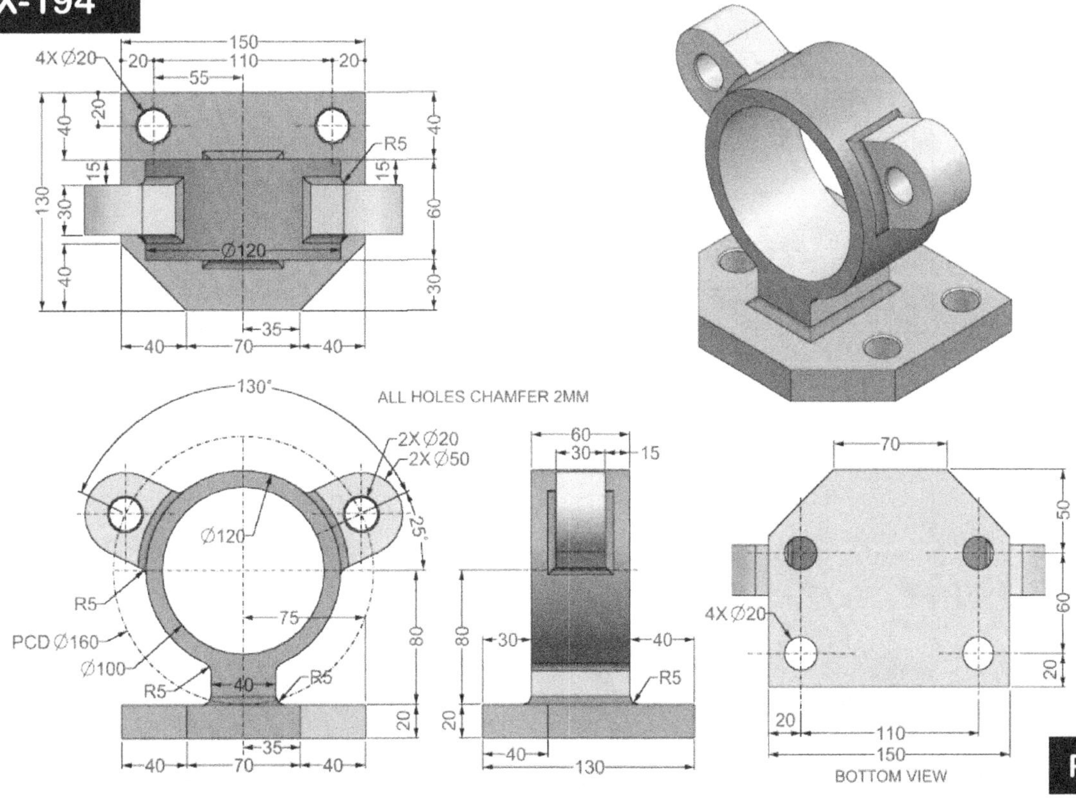

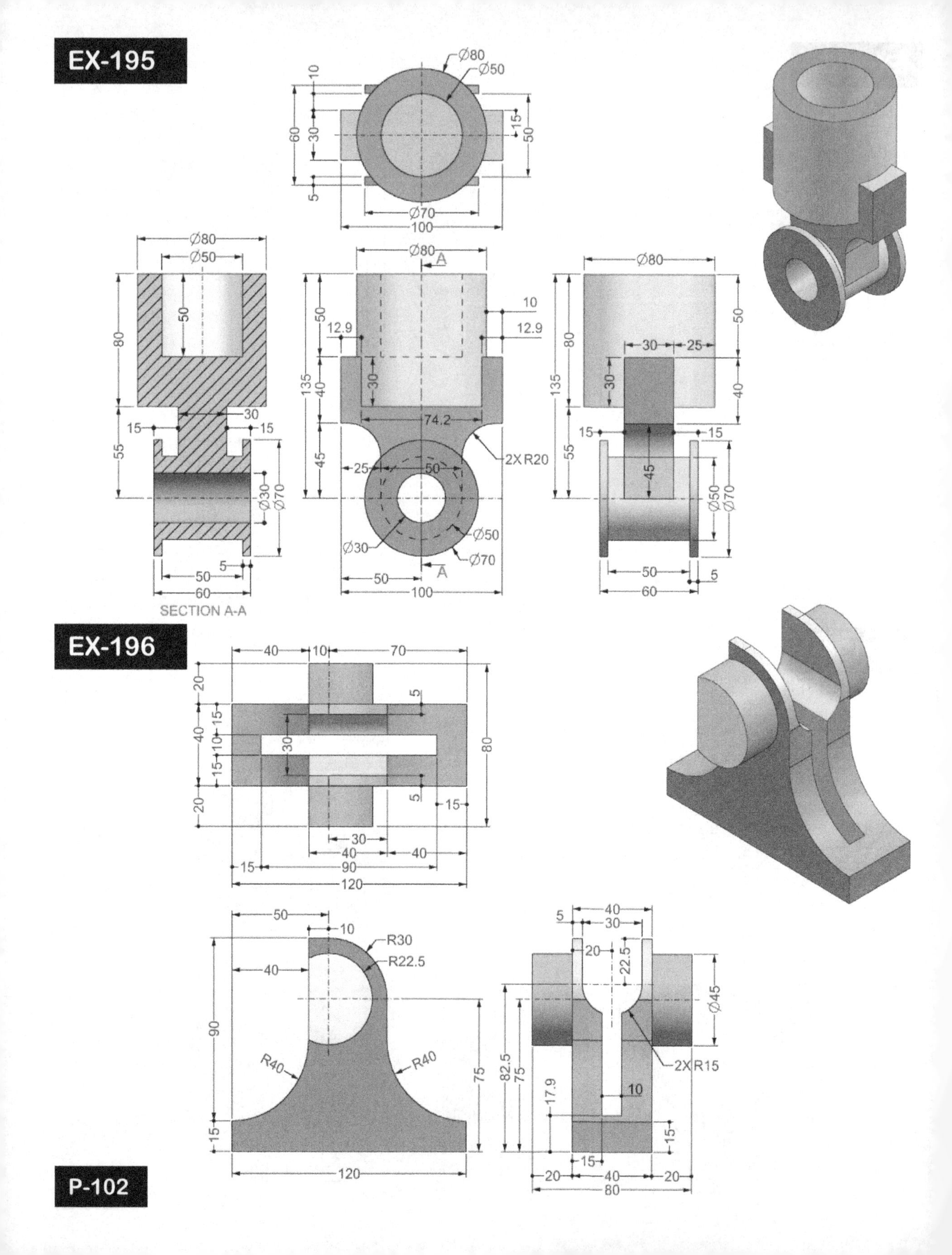
EX-195
Ø80
Ø50
10
60
30
15
50
5
Ø70
100
74.2
12.9
135
40
45
55
80
25
2X R20
Ø30
A
SECTION A-A
EX-196
40
10
70
20
15
5
120
90
R30
R22.5
R40
75
82.5
17.9
22.5
Ø45
2X R15

EX-197

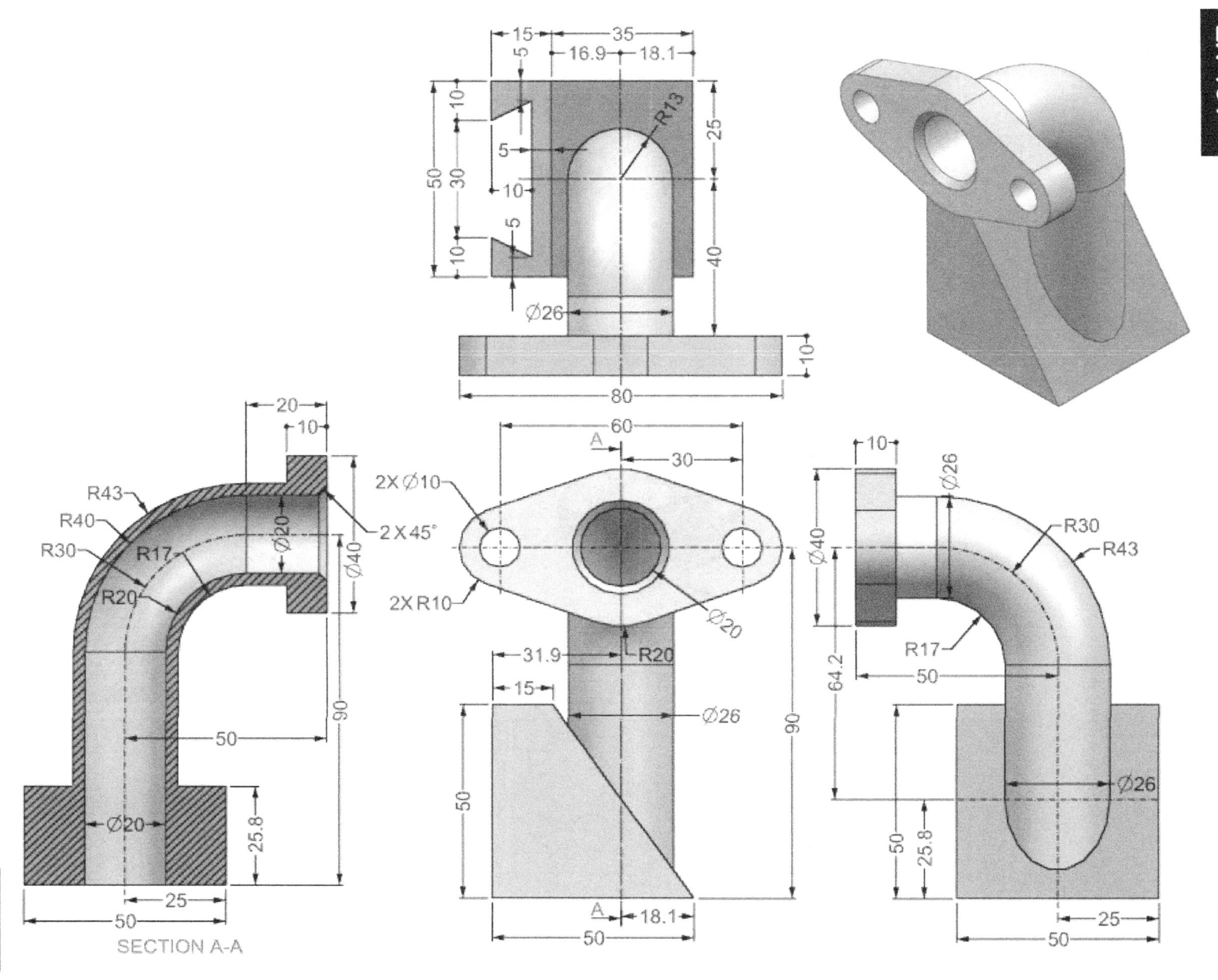

R13
Ø26
Ø20
Ø40
R43
R40
R30
R17
R20
2X 45°
2X Ø10
2X R10
A
SECTION A-A

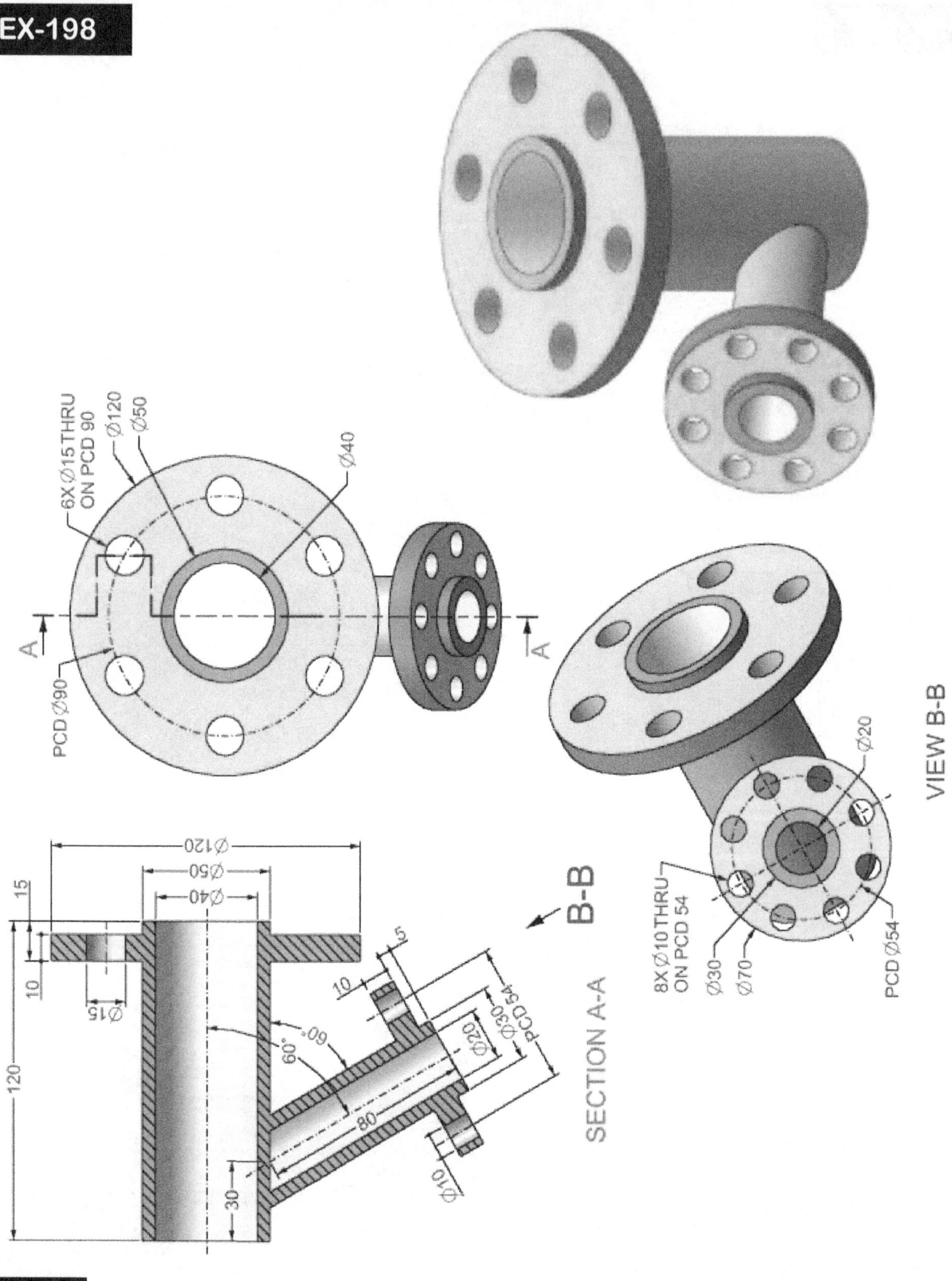

EX-198

SECTION A-A

VIEW B-B

EX-199

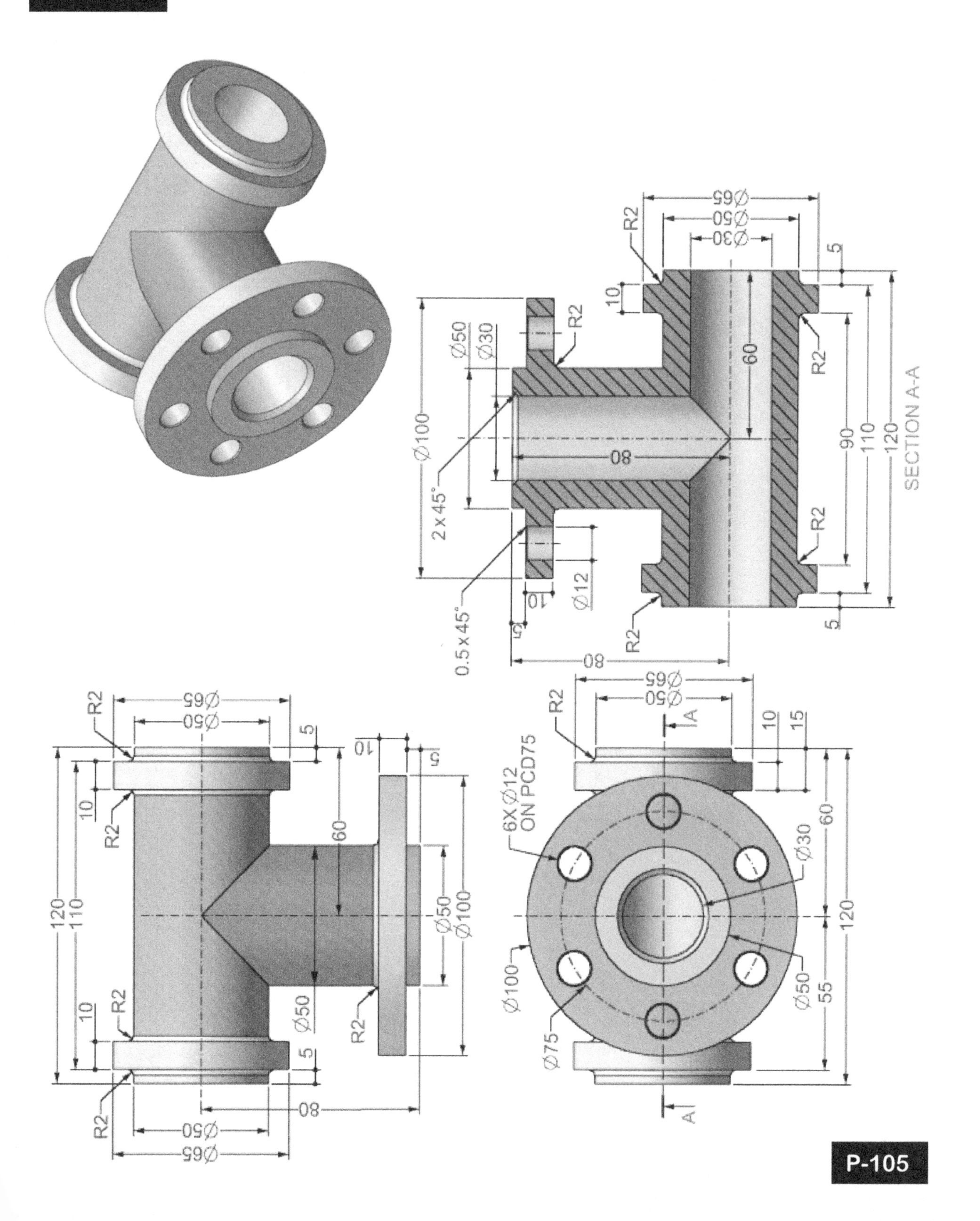
SECTION A-A
6X Ø12
ON PCD75
2 x 45°
0.5 x 45°
Ø100
Ø75
Ø65
Ø50
Ø30
Ø12
R2
120
110
90
80
60
55
15
10
5
A

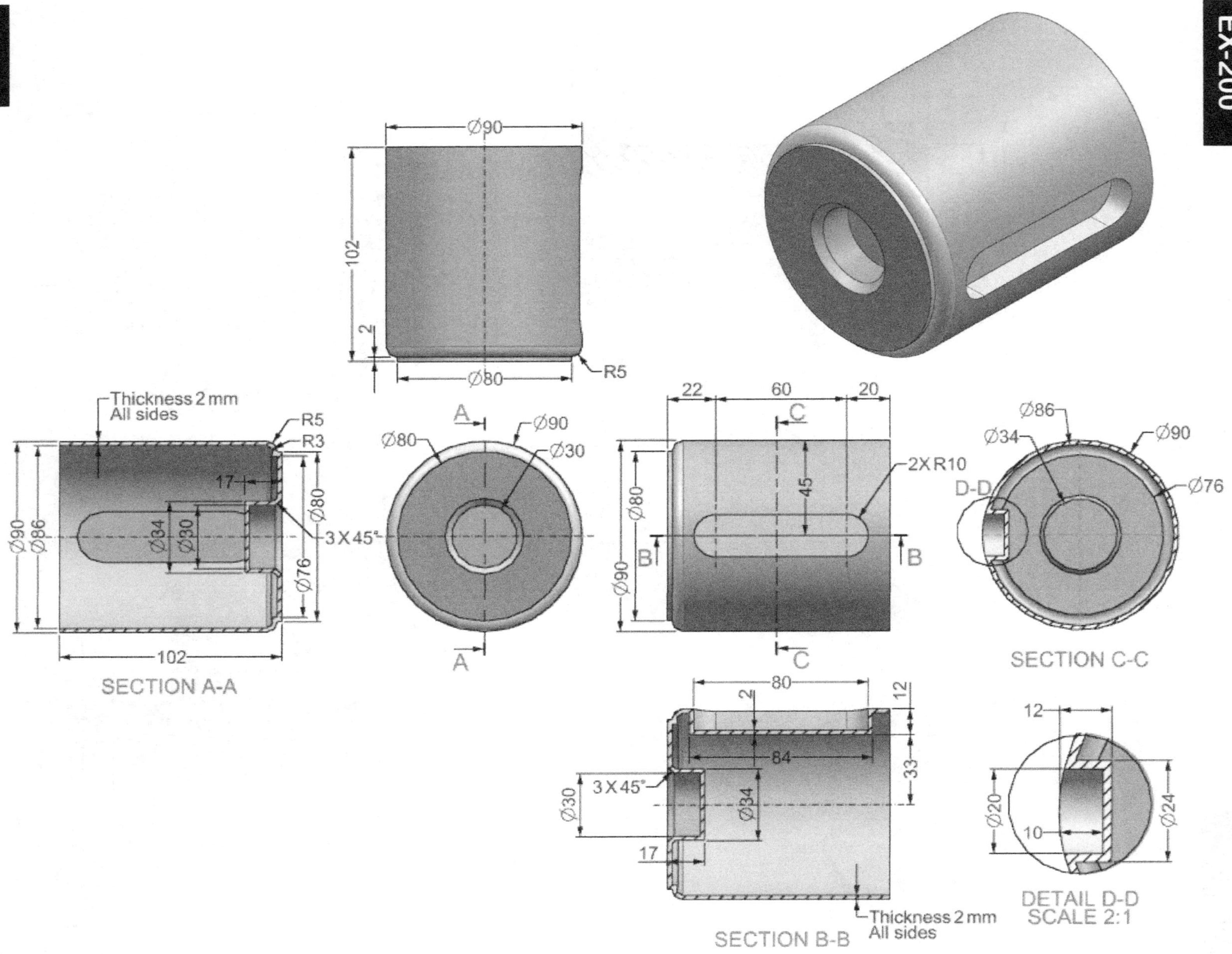
Ø90
102
2
Ø80
R5
Thickness 2 mm
All sides
R5
R3
17
Ø90
Ø86
Ø34
Ø30
Ø80
Ø76
3 X 45°
102
SECTION A-A
A
Ø80
Ø90
Ø30
A
22
60
20
C
45
2X R10
Ø80
Ø90
B
B
C
Ø86
Ø34
Ø90
Ø76
D-D
SECTION C-C
80
2
12
84
33
3 X 45°
Ø30
Ø34
17
Thickness 2 mm
All sides
SECTION B-B
12
Ø20
10
Ø24
DETAIL D-D
SCALE 2:1

Other useful books by CADIN360

1. 150 CAD Exercises

2. AutoCAD Exercises

3. CAD Exercises

4. 50+ SolidWorks Exercises

5. SolidWorks 200 Exercises

6. Autodesk Inventor Exercises

7. Catia Exercises

Made in United States
North Haven, CT
04 July 2024